Stalking the Ghost Bird

The Elusive Ivory-Billed Woodpecker in Louisiana

LOUISIANA STATE UNIVERSITY PRESS BATON ROUGE

STALKING THE GHOST BIRD

Michael K. Steinberg

Published by Louisiana State University Press

Manufactured in the United States of America
First printing

Designer: Michelle A. Neustrom
Typefaces: Whitman, Black Widow
Printer and binder: Thomson-Shore, Inc.

Library of Congress Cataloging-in-Publication Data

Steinberg, Michael K., 1965–
Stalking the ghost bird : the elusive ivory-billed woodpecker in Louisiana / Michael K. Steinberg.
p. cm.
Includes bibliographical references and index.
ISBN 978-0-8071-3305-7 (cloth : alk. paper)
1. Ivory-billed woodpecker—Louisiana. I. Title
QL696.P56 S74 2008
598.7'2—dc22

2007026290

The paper in this book meets the guidelines for permanence and durability of the Committee on Production Guidelines for Book Longevity of the Council on Library Resources. ♾

For my mother and father

Contents

Illustrations

Following page 38

Acknowledgments

Writing this book has been the most enjoyable project of my academic career to date. A few things made this work so enjoyable: first, the fine people that I met along the way; second, the subject of the book and the environment under study (the ivory-billed woodpecker and bottomland forests, respectively); and third, the location—south Louisiana. There are no more interesting cultural and environmental landscapes than in southern Louisiana. While many regions in the United States succumb to a placelessness where everything and everyone seem to look the same, south Louisiana and its people continue to hold on to and celebrate their uniqueness.

I am extremely grateful to the many individuals who assisted me in this writing journey. In Louisiana, "Team Elvis South"—Tommy Michot, Wylie Barrow, Garrie Landry, and Dwight LeBlanc—provided many, many interesting stories about the ivory-bill. Without them pointing me to new leads and hashing over ideas, the book simply would not have developed. Their support and friendship made this project enjoyable. Tommy Michot and Wylie Barrow also kindly let me use several of their photos. Keith Ouchley at the Louisiana Nature Conservancy and James "Van" Remsen at LSU were most helpful in providing information about past sightings and current searches. Van answered numerous questions over the past few years, never seeming to grow impatient with me. Both Van and Keith also provided an insider's view to the current debate about the ivory-bill. Also at LSU, Vern Wright and Bob Hamilton shared their fascinating stories about the ivory-bill and birds in Louisiana in general. Both were very generous with their time and knowledge. Several wildlife and conservation professionals and part-time birders, including Kelby Ouchley, Nancy Higginbotham, Jay Huner, Debbie Fuller, and Chuck Hunter, shared both their personal experiences and their professional insights. All provided valuable opinions about the ivory-bill and bottomland forest conservation.

I owe an enormous debt to the many outdoorspeople in Louisiana who were kind enough to talk with me about their ivory-bill experiences. Their stories and perspectives are central to this book. Fielding Lewis shared many fascinating stories with me not only about the ivory-bill but also about past hunting and fishing adventures in south Louisiana. He was always willing to visit with me when I was in the Franklin area. Similarly, Scott Ramsey was generous with his time and his ivory-bill stories. He allowed me in on the "ground floor" of the search for the ivory-bill in Louisiana, and for that I am most grateful. Also, I would like to thank Jay Boe, Dean Wilson, Betty, and Joe Macaluso at the Baton Rouge *Advocate,* and numerous other sportsmen, some of whom remain anonymous, who told me their stories about ivory-bills, duck hunts, water moccasins, and their love for the Louisiana outdoors. The ivory-bill survives because of the efforts of hunters, fishermen, and conservationists who have protected thousands of acres of bottomland forest over the past fifty years. I would also like to thank Talena Adams, who helped me organize and map past ivory-bill sightings, and Mary Lee Eggart, who produced the final versions of the maps.

I would also like to thank the staff at LSU Press. My acquisitions editor, Joe Powell, and the Press director, MaryKatherine Callaway, were enthusiastic about the book from the beginning. The comments of the LSU Press editorial department as well as freelance editor Susan Brady also made the book stronger. Rand Dotson, also at LSU Press, proved to be a great friend and adviser during the writing and revision phases of this book. We have spent many hours talking about writing and publishing at Baton Rouge's City Park golf course. His advice (about both the book and my golf game) was always levelheaded and positive, which was much appreciated. Outside of LSU Press, my appreciation goes out to editor Christa Frangiamore, who expressed great interest and enthusiasm in the project at the very beginning; Candis LaPrade, whose comments on an early version of the manuscript helped a great deal; friends John and Mindy McCawley, who were always eager for news about the project; and Twain Braden.

I would like to thank my wife, Barbara, for her support of this project. She proved to be a valuable editor and was always enthusiastic about my stories when I returned home. She never wavered in her support of my writing and field research, even when it involved the purchase of critical (and expensive) equipment such as a new canoe. My two sons, Frederick and Truman, also deserve my thanks for their unending excitement and questions about the ivory-bill and its forest home.

Stalking the Ghost Bird

INTRODUCTION

Of all the North American birds, the ivory-billed woodpecker may be the most charismatic. Nicknamed the "Lord God Bird," the "Grail Bird," "King of the Woodpeckers," and "Elvis in Feathers," the ivory-bill—thought to have been extinct since the 1940s—fascinates people from all walks of life and has done so for centuries. The 2004 rediscovery of the bird in Arkansas by a team of Cornell University–led ornithologists has set off a frenzy of media coverage and debates in scientific journals and among birders and ornithologists. This book engages the debate over the bird's current status by examining both the efforts to find this extremely rare species and the reported sightings of the ivory-bill in Louisiana. I use extensive interviews with conservation officials, ornithologists, and natives to illuminate the ongoing controversy and investigate why the ivory-bill, more than any other bird, has excited so much attention and debate.

My own fascination with the ivory-bill goes back at least thirty years and led me to start writing this book a couple of years before the Arkansas recordings were made public in 2005. Some of my earliest memories involve birds. As a boy, I built birdhouses and feeders, placing them around a nearby tract of woods and fields in the hope of attracting bluebirds that had long since departed the St. Louis suburbs. I studied and drew birds using an old hardcover copy of the Golden field guide *Birds of North America*. That field guide stayed with me for years.

I was fortunate enough to spend a number of childhood summers at the Missouri Botanical Garden's arboretum in Gray Summit, Missouri. My great-uncle was the arboretum's caretaker, which allowed us the privilege of staying in one of the houses on the property during my summer vacations. Unlike the woods near my suburban home, these were *real* woods where the boundaries seemed days away; in the late 1960s, they probably were. And while the woods near my home housed only the most common species, the arboretum's woods were home to owls, hawks, and many species of woodpeckers and warblers. Later on, I hunted with my father and learned to identify game birds and waterfowl. Painting decoys is one of the best ways to learn to identify ducks. I liked to take mallard decoys and turn them into canvasbacks, pintails, and scaups. These early experiences instilled in me a deep fondness for woods and fields, rivers and lakes, and especially the animals and fishes that inhabit them.

Given my fascination with birds, perhaps it's not terribly surprising that I would eventually write a book about one of the most famous and even mysterious birds, the ivory-billed woodpecker. But I did not undertake this project as an academic exercise or in response to the Arkansas rediscovery news. I wrote the book simply because I have followed the story of the ivory-billed woodpecker for many years. My fascination with the bird can be traced to a single incident: I thought I saw an ivory-bill when I was a boy.

The sighting occurred while I was vacationing with my family near Ft. Myers, Florida, in the mid-1970s. I noticed three large woodpeckers on a tree in a coastal pine forest. Struck by their size and beauty, I watched the trio closely for some twenty minutes while two of the three birds moved up and down the lower trunk of a pine tree. To this day I remember the calls of the male—a single loud note or honk similar to, but louder than, that of the more diminutive hairy woodpecker. Neither did these single, successional calls resemble the typical high-pitched laugh of the pileated woodpecker. Another point that stood out was that there were three birds together—two involved in the strange dance, bobbing their heads, climbing up and down the trunk of a large pine tree. At the time, I really thought I had seen a family of ivory-bills. I consulted my field guide and read that the ivory-bill was near extinction, but I doubt I understood what that meant. Upon returning home, I sat down and wrote the National Audubon Society, drawing and writing a description of what I saw. I never received a response.

Even though I realize now that what I saw was probably a pair of pileated woodpeckers involved in a courtship ritual, since that childhood experience in Florida I have always followed the quest for the ivory-billed woodpecker with keen interest. Over the years, I have read all of Jerome Jackson's articles on the bird; I even have an old, tattered copy of James Tanner's published dissertation on the ivory-bill. And although my lifelong interest in birds and bird-watching did not result from that experience alone, it nonetheless inspired me at a relatively early age to learn as much as I could about a rare species. While my own research and writing as a geography professor have usually taken me to the Central American tropics—far from the ivory-bill's habitat—I remain drawn to the flora and fauna of the southern bottomland forests, which contain some of the most diverse, unique environments outside the tropics.

While my original plan for researching this book was to investigate the possible presence of ivory-bills and the stories associated with the bird throughout the South, my focus narrowed because so many of the recent reported sightings were in Louisiana. Most birding field guides list South Carolina as one of the last refuges of the ivory-bill, and the quality of the habitat in places like the Congaree Swamp certainly supports that state as a site for future ivory-bill searches. The Big Thicket area in east Texas has also produced many solid sightings, as reported by naturalist John Dennis into the 1970s (Dennis 1967, 1988). But as much as I would have liked to look for the bird and talk with local people in South Carolina and Texas, there were just too many interesting stories coming from Louisiana to ignore. While this book does not focus exclusively on Louisiana, most of these stories are set in the Bayou State. Similarly, while there are many people—southerners and nonsoutherners, renowned scientists and nonprofessionals—who play a role in the ivory-bill's story, I have chosen to focus on the many people in Louisiana who are fascinated with and have insights about the ivory-bill. From Audubon's writings about the St. Francisville area to Tanner's study of the ivory-bill in the late 1930s in northeast Louisiana, the history and lore of the ivory-bill are interwoven with Louisiana and its citizens.

This book presents the stories of both experts and local people in Louisiana. The stories document their encounters with and insights about the ivory-bill and testify to a shared fascination with the bird. This book also presents information on a 2005 search for the ivory-bill in Louisiana that, while it did not produce any photographs, did result in a couple of sightings by experienced birders. You, the reader, can choose to believe the stories of sightings detailed in this book or dismiss them. But what struck me as I collected these stories is that there are simply too many knowledgeable people in Louisiana who have reported seeing the ivory-bill to dismiss them all.

Chapter 1 provides background information on various ivory-bill topics such as the ongoing controversy about the 2004 Arkansas rediscovery, the historical fascination with the bird, and the alteration of its habitat. This chapter also provides information about the 2005–6 search for the ivory-bill in Louisiana organized by the Louisiana Nature Conservancy. Chapter 2 explores in more detail the human impacts on both the ivory-bill and its bottomland forest habitat. It also examines forest-management practices in the South, and how current practices may pose problems for an ivory-bill recovery. In chapter 3, I interview various conservation and ornithological experts. Here experts relay their thoughts about the ivory-bill's status and its prospects for recovery. A few of the individuals interviewed in this chapter have had encounters with the ivory-bill themselves. The observations they offer are not simply those of "outsiders"; these individuals have deep personal interest in the ivory-bill's plight. Chapter 4 turns to the stories of local people. Louisiana offers a rich cultural landscape, and the stories from the local people about their encounters and fascination with the ivory-bill reflect the interest and variety of that landscape. Reports of ivory-bills made by rural people, hunters, and fishermen are often dismissed by birders and ornithologists. I argue that greater attention should be given to these stories, which often merit investigation. It is the outdoors folk who spend the most time in the vast ivory-bill habitat, which is often remote and difficult to access. In chapter 5, I outline where I believe future searches should take place based on current habitat quality, number of past sightings, and the opinions of various conservation and ornithology officials. Appendix 1 provides an overview of the locations of ivory-bill sightings, decade by decade, starting in the 1950s. Appendix 2 consists of a more detailed timeline of reported ivory-bill sightings and their specific locations.

A bibliographic essay precedes the reference list and provides a brief discussion of some of the most informative and useful works on the ivory-bill and its environs.

1

BACKGROUND

Among birders and ornithologists, especially in the South, the ivory-billed woodpecker (*Campephilus principalis*) is a haunting presence whose existence is debated at meetings, in articles, and on the Internet. Thought to be extinct since the 1940s, the ivory-bill is the largest woodpecker in the United States. Its size, along with its distinctive top crest, coal black feathers, ivory bill, and yellow eyes, inspired the people who once lived near it to nickname it the "Lord God Bird"—the exclamation its unexpected sighting was thought likely to evoke from a startled observer. The bird's striking appearance conjures up images of deep and foreboding bottomland forests, of Audubon exploring and painting the South, and of a wild southern landscape home to wolves, panthers, and innumerable species of birds, long before the southern forests were felled to create fields for cotton and soybeans, and long before the region's rivers were tamed to control floods.

Thus, when it was announced on April 28, 2005, that the ivory-billed woodpecker had been rediscovered in Arkansas, the birding, ornithological, and conservation communities lit up with excitement and interest. This astonishing announcement was the result of a yearlong, intensive search in the Cache River and nearby White River national wildlife refuges in Arkansas organized by the Cornell Laboratory of Ornithology, as well as the Nature Conservancy's Arkansas chapter and various individuals who had ornithological and birding credentials (herein to be described as the Cornell-led search team) (Fitzpatrick et al. 2005).

The ivory-bill's status had been debated for more than sixty years prior to the Arkansas report (Jackson 2002, 2004). Had the bird survived? If so, where was it? The authenticity of the few photos of the bird that surfaced previous to the Arkansas sightings continues to be disputed, sometimes heatedly, by the ornithological and birding communities, as are the many reported sightings from that time. Some of these sightings have seemed credible enough to make even hardened skeptics think twice about the ivory-bill's status (see appendix 2; Baker 1950; Dennis 1967; Hamilton 1975; Lewis 1988). But even after the Arkansas sightings and eventual brief filming, consensus remains elusive.

Why does the story of the ivory-bill fascinate? Why has it claimed space on the front pages of the *New York Times* and the *Wall Street Journal*? Why is its status debated in the pages of *Science*? Is it the beauty and size of the bird? Does the ivory-bill, like bottle trees and juke joints, strike outsiders as a unique and fascinating feature of the southern landscape? Or is it more than simply the bird's physical uniqueness? Is it the idea that the ivory-bill has seemed to rise from the dead—that a species could reappear years after being labeled extinct—that people find so compelling? The media frenzy created by the rediscovery seems to bear out the appeal of this idea to both the scientific community and mainstream society.

The intrigue surrounding the bird and its sporadic appearances has bestowed an almost mythic identity on the ivory-bill. It has become a ghost bird, appearing and vanishing, offering no more than glimpses and leaving nothing but stories. Because decades in the twentieth century passed without any widely accepted sightings, leading ornithologists assumed the bird was extinct. However, there remained the faithful who, for one reason or another, held out hope that the ivory-billed woodpecker would appear again. That hope was sustained by sporadic reports of ivory-bill sightings in deep southern swamps. And then, in almost Hollywood fashion, the faithful were rewarded with the announcement in 2005 of the Arkansas team's findings. The team's film clip was ever so brief, leading some skeptics to question its validity. The Cornell-led team also had made multiple sightings and audio identifications of the ivory-bill, however, and the apparent authenticity of the audio recordings prompted two ornithologists who were initially skeptical of the rediscovery to withdraw a paper they were preparing that presented their concerns (see the 2005 *New York Times* article by Gorman and Revkin).

Controversy and hardened stances over the bird's status are not new chapters in the ivory-bill story. For decades, there have been individuals within the ornithological and birding worlds who vehemently asserted the bird existed, while others scoffed at such an idea (Dennis 1967; Short 1987; Cokinos 2000; Weidensaul 2002).

Even though the evidence produced by the Arkansas search team has been met with skepticism in some quarters of the birding and ornithological worlds, the research I conducted while writing this book—which began well before the Arkansas team released its findings—has left me with little doubt that the ivory-bill survives. As I began this project, the question that spoke to me was, Why aren't the stories of the ivory-bill going away? And my subsequent research has led me to too many knowledgeable people who are certain that they have seen an ivory-billed woodpecker—a few on multiple occasions—for me to believe that the bird is extinct. I had not initially intended to take such a definitive stance. But my encounters with these credible eyewitnesses began to change the course of the book I would write. I began to document the stories of these Louisiana outdoorspeople and to explore why they believe the bird survives. I also spoke with individuals in Louisiana who believe the ivory-bill is gone forever. When the bird's rediscovery was announced in 2005, my project suddenly shifted again. The people I had been interviewing who held out hope that the ivory-bill survives were vindicated (at least in the eyes of many), and I returned to document their reactions to the rediscovery. The stories of these Louisianans, collected herein, are central to this project.

These eyewitness accounts relate to an important theme in this book—the proprietorship of rediscovery. In other words, who determines when something is rediscovered? Is it a group of elite birders with expansive life lists? Is it the federal government, the Nature Conservancy? Or can local people or state wildlife officials be credited with rediscovery?

Rediscovery must be sanctioned by a scientific or government body at some point and at some level. But to many of the individuals featured in this volume, rediscovery took place well before the Arkansas announcement. Their proof is their word, their memory, their honor. For these outdoorspeople, some of whom are unaware of the raging debate in the birding and ornithological worlds, there is no question that the ivory-bill survives. And they are sometimes indifferent to whether outsiders believe their stories.

The birding world, on the other hand, tends to be competitive, with rediscovery the ultimate achievement for a member of the birding tribe. The quest to add rare birds to their "life list" drives many competitive birders, who spend a great deal of time and money on this effort, often visiting far-flung places, and who scrupulously evaluate claims of unusual sightings by others. There also exists a distinct hierarchy in the birding world, and reports of rare birds made by hunters and fishermen appear to rank low in the estimation of many in the birding caste system. The distrust of local people—people outside the ranks of the competitive birding world—has kept the ivory-bill hidden longer than it should have been.

Now that the ivory-bill has been found (albeit fleetingly), perhaps we are entering a new phase of ivory-bill searches with more cooperation between the birding and scientific establishments and the local people, who are often more knowledgeable about the local landscape than outsiders. A search that began in Louisiana in the summer of 2005 was based on a reported sighting by a hunter, while new searches in South Carolina and Texas are also taking place at sites where local people have sporadically reported the ivory-bill for many decades (Dennis 1967; Jackson 2004).

But for now, with the ivory-bill story still mired in professional politics, it may be that at least part of the bird's story could be better told by someone outside the ornithological and birding worlds, someone who can shed light on the human dimension of the quest to find and save the ivory-bill. Being a geographer—someone who studies human-environmental interactions and human impacts on the environment—rather than an ornithologist has influenced how I look at birds in general and the ivory-bill specifically. I fall outside the hierarchy of both the birding and ornithological worlds. I am fascinated with birds; I watch birds. I have even led birding tours in Costa Rica. But I see birds as part of a larger landscape, with presence and status in that landscape greatly influenced by history, politics, economics, and, of course, local culture. The ivory-bill, more than any other bird at this moment, is influenced by these factors. Setting aside land is expensive, and buying private land for conservation purposes is inherently political. For me, one of the most critical and interesting aspects of the ivory-bill story is that very credible reports of the bird by local people who hunt and fish (insiders) are disputed by experts (outsiders). From a geographer's perspective, this might suggest that the birding and ornithological communities would benefit from appreciating and incorporating a more geographical, cultural-ecological perspective in future research concerning not only the ivory-bill but also other rare species that are impacted and influenced by human decisions and cultures.

I also am interested, as a geographer, in how people relate to, value, and perceive nature. Birds have long fascinated human culture. The power of flight, the use of feathers in art, rituals, and fashion, and the symbolism of birds provide insight into how birds are valued by human cultures (Doughty 1975; Bonta 2003). Although some today may think only primitive cultures in remote lands continue to value the symbolic and spiritual qualities of birds, no one can deny that the ivory-bill is a powerful symbol today—a spiritual symbol to some—representing a more environmentally pristine South. What follows is my attempt to tell the cultural-ecological story of the ivory-bill in Louisiana, and the story and issues surrounding its recent rediscovery.

The Rediscovery in Arkansas

The Arkansas search was initiated after a kayaker, Gene Sparling, saw a large, crested woodpecker in the Cache River National Wildlife Refuge on February 11, 2004. Sparling reported the sighting and a description of the bird on an Internet Web page, also mentioning that it was a woodpecker he had never seen before. The Web report drew the interest of Tim Gallagher, the director of publications at the Cornell Laboratory of Ornithology, and Professor Bobby Harrison of Oakwood College, individuals who were already fascinated with the ivory-bill. In the case of Harrison, it was an outright obsession. Harrison had intermittently searched for the ivory-bill for thirty years. This initial report, coupled with a later sighting on February 24 by Gallagher and Harrison in the same area as the first one, set in motion the single most intensive search for the ivory-billed woodpecker. The search was led by Professor John Fitzpatrick of Cornell University, along with the Nature Conservancy of Arkansas, the U.S. Fish and Wildlife Service, other individuals associated with Cornell's Laboratory of Ornithology such as woodpecker expert Martjan Lammertink, and a host of volunteers. Five more sightings were made between April 5, 2004, and February 15, 2005.

On April 25, 2004, David Luneau, a search team member, caught an ivory-bill on film as it flew away from a kayak. This four-second video has been at the center of the recent controversy regarding the validity of the Arkansas sightings. Overall, the search team logged more than twenty-two thousand hours of search time between February 2004 and April 2005 (see Gallagher 2005; and Fitzpatrick et al. 2006 for a detailed account of the Arkansas search).

The Louisiana Nature Conservancy's 2005–6 Search

In addition to the Arkansas search and sightings by the Cornell-led team, in 2005–6 the Louisiana Nature Conservancy directed an intensive field search in far southern Louisiana after hunter and local landowner Scott Ramsey gave a detailed description of one or more ivory-bills he had sighted near Patterson (map 1). The scale of the search in Louisiana was small compared with the one that took place in Arkansas, but initially at least, the search found tantalizing evidence that the ivory-bill also survives in southern Louisiana. Given the number and consistency of reported sightings from that area over the past few decades, it may ultimately be in southern Louisiana—rather than in the White and Cache river region of Arkansas, where national attention has lately been focused—where breeding pairs of ivory-bills are located.

Before the 2005–6 Louisiana search, the general Patterson area was already believed to have potential as an ivory-bill habitat because of numerous reports of past sightings (map 4). The Louisiana State University (LSU) ornithologist George Lowery mentioned the area in his *Louisiana Birds* after he was sent several photos of an ivory-bill in 1971 that were taken nearby (Lowery 1974). The controversial photos were taken by Fielding Lewis, a resident of Franklin, while training his retrievers in a nearby swamp (figure 1). However, it was not until a chance meeting in 2004 between landowner Scott Ramsey and birder Jay Huner that the area became an official focal point in the search for the ivory-bill.

In the spring of 2004, Jay Huner, the retired director of the Crawfish Research Center at the University of Louisiana (formerly known as the University of Southwestern Louisiana), was collecting bird diversity data near Patterson for a study unrelated to the ivory-bill. Ramsey spotted Huner on the side of the road near his property, prompting an inquiry. After Huner explained to Ramsey what he was doing, Ramsey invited Huner to come to his property and take a tour of some of the trails and habitat. Huner had mentioned that ivory-bills had been reported in the region, but Ramsey was not familiar with the bird. He knew the property housed large woodpeckers, but he hadn't given the specific species much thought.

The following winter (around February 2005), Ramsey contacted Huner at his office at the University of Louisiana in Lafayette, telling Huner that he had seen what he believed to be both male and female ivory-billed woodpeckers. Ramsey said that he had seen several large woodpeckers, bigger than the pileated, that were sort of "goofy" in appearance. (This description gave rise to the code name "Goofy" that was used for the ivory-bill in the Louisiana search.) Ramsey asked Huner to come to Patterson to see if he could confirm the identification, but Huner was not able to return to Patterson until early May. Although Huner did not see an ivory-bill when he did visit, he did believe he heard an ivory-bill "tooting" (Huner's description) on the property.

Huner and Ramsey then made contact with the USGS National Wetlands Research Center in Lafayette and the U.S. Fish and Wildlife Service to alert them to the presence of ivory-bills in the area. At about the same time, various other individuals in the state had heard the rumors swirling about sightings of ivory-bills in different locations in Louisiana (including those based on the reports of many of the people featured in this book), but the past and recent sightings in the area between Patterson and Franklin, Louisiana, were causing mounting interest in a search of that area. Those who wanted to follow up on these reports eventually fell under the direction of Keith Ouchley at the Louisiana Nature Conservancy, who agreed to sponsor the Coastal Forests Wildlife Inventory project. This rough timeline of ivory-bill search events was relayed to me by Jay Huner and Keith Ouchley.

The search initially started out promisingly. Huner believed he had heard an ivory-bill in May, and late in the evening of July 7, he thought he, along with Ramsey and his partner, saw a male ivory-bill. Below is the e-mail correspondence provided by Huner to the search group and also sent to me:

> 8:30 PM CDST July 7, 2005
>
> In the absence of "proper paper," this will have to do. Today, at approximately 7:40 PM CDST Scott Ramsey, partner, and Jay Huner (me) saw a very large woodpecker on the northwest side of the No Name Hunting Club property adjacent to large borrow pits. Light

was bad although the sky was basically clear at the time. The bird landed at the top of a 40–50' tall tree about 2–2½' in diameter at the base (species to be ascertained). The upper third of the tree was dead. I did not see the bird land. Partner saw it and we all turned to look. It dropped down quickly and flew to a vine-covered cypress about 40 yards behind the tree in which it landed. Partner had good views, but neither Scott nor I saw it well. Partner saw it in the second tree, but by the time Scott and I saw it, it had flown further back to several cypress trees another 50–60 yards beyond and away from us. It was simply too dark to even consider trying to shoot images with a digital camera.

I subsequently saw the bird fly in an arc to the south and was able to get reasonable views of it in flight as it crossed behind the trees and passed through two gaps in the woods.

Body notes: (1) Very big for Pileated Woodpecker.

(2) Very prominent crest (Red)

(3) Very strong, straight flight, not undulating like Pileated Woodpecker

(4) white-underside of wing / not spot [*sic*] like Pileated Woodpecker

[Note: Diagrams of crest and wing patterns on original notes.]

[Note: Diagram of locations where bird flew and seen on original notes.]

I saw the bird in flight only and could not really focus to pick up on the expected white of an Ivory-billed Woodpecker but was impressed by the amount of white I did see as the bird crossed the second gap.

I have previously encountered Pileated Woodpeckers at the site at least 3 times since late May and as recently as 1 July of this year. This is the site where Partner & Ramsey have concluded that they have seen at least one Ivory-billed Woodpecker since February of this year.

Jay V. Huner [signature]

PS—I believe that the collective observations show that this was a male Ivory-billed Woodpecker.

In addition to Huner's reported sighting, another search team member, who wished to remain anonymous, also had one particularly good sighting. Below is his exact field entry of his encounter with an ivory-bill. He transcribed his notes and sent them to me via e-mail.

Date: August 1, 2005.
Time: 3:25 pm.
Location: Approximately 150 yards north of gate to landowner's property. Sighting of target species—distance paced off was ~40 yards from location of 1st sighting of bird. Bird was on the east side of Said Road apparently trying to fly across the road to the west side toward area # 1. I observed a woodpecker larger than the Pileated Woodpecker fly out of the trees lining the east side of Said Road. As it flew towards me, it evidently spotted me, flared and rapidly flew into a nearby sweet gum tree.
Identification: I was not using binoculars. The entire initial sighting lasted about three (3) seconds. The bird seemed to be about ⅓ larger than the Pileated Woodpecker. What caught my attention was the entire trailing edges of wings were a very white coloration (wingtips looked dark). The forewings were dark. The bird's body appeared to be a black that was somewhat brighter black than the Pileated Woodpecker. When it apparently spotted me near its intended path of flight, it quickly flared and retreated back to the area from which it had appeared. It only made it to a few feet from the power lines before it retreated.

In the brief sighting, I could not discern if it was male or female although in hindsight, I feel I saw some red on its head area. The sighting was just too brief to be definitive on its sex. I did not discern its bill color since the most attention-getting field mark was the white and black wings flashing. Once the sighting was ended, I heard a loud rap, rap-rap, rap, rap-rap, etc. coming from a sweet gum tree about 20 yards from the initial sighting. This lasted ~30 seconds. I carefully tried to go through the thick vegetation but it was difficult. I finally managed to penetrate it and searched the trunk of the sweet gum tree. I could not locate the bird although it was hitched to that tree. A few minutes later, from that tree, I briefly saw the silhouette (no colors at all) of an extremely large woodpecker rapidly flying away from that tree. A few seconds later, I again heard the same rapping coming from a tree that I estimated was about 100 yards into the woods. After a couple of minutes, the rapping stopped and the bird apparently departed. I searched that area of the woods for about ½ hour. I am 100% sure of the field marks I've reported. Field notes were made within five (5) minutes of sighting.

Field guide was not consulted in writing this description.
Anonymous [signature]

These are the types of reports that convinced me that the ivory-bill survives in Louisiana. Neither Huner nor the anonymous team member is a novice birder. Huner has been a birder for many years, sometimes as a contract consultant on birding surveys. The anonymous team member lives in New Iberia, a town on the edge of the Atchafalaya Basin. A birder for more than forty years, he also has worked as a contract ornithologist conducting bird surveys on rice and crawfish farms for more than ten years. Their experiences make it extremely unlikely that they would have confused the birds they saw with any other bird, even the similar-looking pileated woodpecker.

Unfortunately, the Louisiana search was interrupted when Hurricane Katrina hit the Gulf Coast on August 29, 2005. The camp in which the search team was staying was needed for evacuees from New Orleans, and the search team had family commitments to attend to related to the storm. Also, most state agencies in Louisiana directed their personnel and resources to deal with the crisis (which was compounded in September by Hurricane Rita, which destroyed much of infrastructure in the far southwestern part of the state). When the search was reinitiated in the spring of 2006, no ivory-bills were sighted or heard. The timing of the hurricane was unfortunate because the search team felt that ivory-bills had been using the area, and that a photo and another rediscovery announcement had not been far off. While the Patterson area did not experience the brunt of the hurricane damage, there may have been enough of a disturbance to drive the birds out of the area, possibly north into the heart of the Atchafalaya Basin. Then, on March 14, 2006, a helicopter crashed near the search location, killing two people and setting ablaze parts of the forest in which the search took place. In total, a couple of hundred acres burned, according to landowner Ramsey. So this area has undergone two significant natural events in the past few years with the potential to drive out any ivory-bills in the area.

Although it is quite disappointing that the 2005–6 Louisiana search produced no photos, Ramsey's sighting did generate renewed interest in finding the ivory-bill among various groups and individuals in Louisiana capable of investing the resources likely needed to find such a rare species. A task force led by the Louisiana Nature Conservancy has been formed to select new search areas in Louisiana based on recent reports of the ivory-bill, with the southern Atchafalaya Basin being one of the focal areas for new searches (figure 2). While no new formal searches had begun as of summer 2007, Ramsey reportedly has occasionally seen and heard ivory-bills on his property in early 2007, while birder Jay Huner believes he saw a single bird in the same area in June 2006 (soon after the formal search ended). In a March 2007 e-mail, Ramsey told me that he does indeed believe the birds have returned to the area. In many ways, their return makes sense. Since Hurricane Katrina damaged and felled trees in the area, the ivory-bills may be returning to feed on the beetle grubs that infest dead and dying trees.

How Many Birds and Where?

In the rush to dismiss sightings and relegate the ivory-bill to the history books, some skeptics of recent searches and sightings have failed to acknowledge how little potential ivory-bill territory has been systematically searched. As of summer 2006, searches for the ivory-bill had taken place in Arkansas, Louisiana, and Florida. But these searches covered a small part of potential bottomland habitat. Other searches in east Texas and South Carolina were then beginning to take shape. Even within these areas, a relatively small fraction of territory was formally searched. Overall, the area surveyed may amount to from 1 to 3 percent of potential ivory-bill territory. Actually slogging through bottomland forests to find this reclusive bird is difficult, and the time, money, and manpower for intensive field expeditions are in short supply. But the simple fact that very little of the bird's potential habitat has been thoroughly searched would seem to make any pronouncement about its extinction premature.

While a handful of locations around the South have the potential to house ivory-bills, the number of excellent reports of sightings coming from Louisiana in recent years makes that state a logical place to expand and intensify searches. Louisiana contains some of the wildest and least explored tracts of wilderness left in the southern United States. Even though these forests may be located in close proximity to human populations (the Pearl River Wildlife Management Area, for example, is close to New Orleans), they are difficult to penetrate because of the seasonal flooding, prickly palmettos, bugs, snakes, mud, and alligators. These bottomland forests are normally not destinations for casual day-hikes or picnics. Until recently, few of the people who entered these landscapes were looking for woodpeckers.

Before the sighting and filming of the ivory-bill in the Cache River National Wildlife Refuge in Arkansas, many experts dismissed this area, along with the nearby White River National Wildlife Refuge, as unlikely locations to find the bird. Both the father of ivory-bill research, James Tanner, and, more recently, woodpecker expert Jerome Jackson have claimed that because of past logging impacts, the habitat contained too little mature forest to support ivory-bills (Tanner 1942a, 1942b; Jackson 1989, 2004). For example, Jackson (2004:158) stated that "superficial examination of habitats revealed few areas of extensive, mature bottomland hardwoods." Much earlier, James Tanner, who spent a total of eight days in the White River Waterfowl Refuge (now a national wildlife refuge) in 1938, found "no indications of the birds still being there." He stated that "there are a few virgin tracts of sweet gum and oak timber but too small and scattered to make really good Ivory-bill territory" (Tanner 1942a:25).

According to the Nature Conservancy, the bottomland forests in the Big Woods area of Arkansas contain about 550,000 acres. This is the largest corridor of bottomland hardwood forest remaining in the Mississippi River Delta north of the Atchafalaya Basin (Nature Conservancy 2006). The term "Big Woods" refers to the floodplain forests lining the Mississippi, White, and Lower Arkansas rivers, and the Cache River in Arkansas. Perhaps more in-depth analyses of the Arkansas habitat would have revealed that this area actually contains some of the best and most expansive habitat in the Mississippi River alluvial valley. Most of these areas were logged at some point over the past one hundred years, but the Big Woods area also contains some decent tracts of old-growth forest. I point this out not to disparage Tanner's and Jackson's work or opinions, but to raise the important issue of how much time, prior to the sighting in 2004, was spent in the area examining habitat or looking for the ivory-bill. To follow a week or so spent in an area with the declaration that the bird is absent is less scientifically valid than the evidence provided by the Arkansas search team in support of its rediscovery. Another point that skeptics may have missed or misunderstand is that the bird or birds sighted in Arkansas may be moving long distances through forested corridors, not just in Arkansas but to forests in neighboring Mississippi, Tennessee, and Louisiana. Thus, while the ivory-bills may spend even large blocks of time in Arkansas, they may not be continuous residents. The White River region is connected via riverine or batture forest corridors to other areas in the Lower Mississippi alluvial valley (map 2). We simply do not know how much territory is used by ivory-bills today.

Historical Fascination with the Ivory-Bill

Centuries before the Arkansas filming of the ivory-bill made front-page news across the United States, humans were fascinated with this huge woodpecker. Many Native American tribes valued its plumage and large bills. Tribes in the southern United States used ivory-bill skins to carry medicine bundles (Jackson 2004). Ivory-bill remains were traded for by Native tribes as far north as Canada and as far west as Colorado, far from the range of living birds (Audubon 1834; A. M. Bailey 1939; Parmalee 1958; Jackson 2004). Early European American naturalists such as John James Audubon, Mark Catesby, John Abbot, and Alexander Wilson also became enamored with the ivory-billed woodpecker, commenting on its stunning appearance and strength. In his classic five-volume *Ornithological Biography*, published from 1831 to 1839, Audubon (1999:269) compares the ivory-bill to the works of the seventeenth-century Flemish portrait and religious painter Anthony Vandyke:

> I have always imagined, that in the plumage of the beautiful Ivory-billed Woodpecker, there is something very closely [*sic*] to the style of the colouring of the great Vandyke. . . . So strongly indeed have these thoughts become ingrafted in my mind, as I gradually obtained a more intimate acquaintance with the Ivory-billed Woodpecker, that whenever I have observed one of these birds flying from one tree to another, I have mentally exclaimed, "There goes a Vandyke!"

Naturalist Alexander Wilson, considered by some to be the father of American ornithology, wrote about the ivory-bill's strength and spirit in his multivolume *American Ornithology*, published from 1808 to 1814 (Wilson 1811: 20–26). After having shot and injured a specimen, he carried it to his room so that he could draw it. In Wilson's absence, the bird nearly destroyed the room. Plaster from the ceiling and walls, and wood chips from the furniture were strewn about the room, the results of the ivory-bill's desperate escape attempt. Sadly, after three days of suffering in Wilson's company, the bird died.

In the twentieth century, the southern literary giant William Faulkner mentions the "Lord God Bird" in his 1942 story "The Bear." Faulkner associates the bird with the southern primordial environment that was quickly disappearing in Mississippi at the time of his writing. While the term "Lord God Bird" also has been used to describe the pileated woodpecker (*Dryocopus pileatus*), the theme in Faulkner's story of a vanishing southern wilderness makes it likely that he was referring to the increasingly rare ivory-bill. The ivory-bill is also mentioned in Walker Percy's 1971 novel *Love in the Ruins*, set in and around Slidell, Louisiana. Percy predates the recent excitement with a small section in his book describing a sighting of the ivory-bill near Honey Island Swamp.

Since the 2005 report of the Arkansas sighting, fascination with the ivory-bill has spread to the greater public. In Arkansas, near the White River region where the bird was spotted, a burgeoning tourism industry has developed focused on the ivory-bill. In Louisiana, I have been surprised by how many people mentioned the Arkansas sighting to me, along with their own story related to woodpeckers or birds in general. In many cases, I was not the one to initiate these conversations; the speakers were often unaware that I was writing a book about the woodpecker, indicating a genuine enthusiasm for the bird. A couple of people I met were in the process of creating and naming a drink in honor of the bird, quite an honor in southern Louisiana. The excitement about the ivory-bill in the Baton Rouge area, and probably elsewhere, has led weekend birders to undertake forays into the bayous and swamps, initiating their own searches for the ivory-bill.

But the public's emotional reaction to the rediscovery of the ivory-billed woodpecker is not unalloyed euphoria; there is also sadness that the bird is so perilously close to extinction. Even if a small number remain, possibly in isolated populations, the ivory-bill is still critically endangered simply because so few birds exist. Therefore the ivory-bill remains a potent symbol of the destruction of the wild South. The largest woodpecker in the United States is absent from almost all of its previous bottomland haunts in the South. It is reminiscent of other species such as the eastern timber wolf (*Canis lupus lycaon*) that require large amounts of territory or are specialists whose ranges declined precipitously as people moved in and their often destructive land-use practices spread west. But the Arkansas sightings and reports from Louisiana bear witness to the fact that this bird is more resilient than previously thought by most. If the ivory-bill has survived this long, perhaps the species has a chance to recover.

No Ordinary Bird

Highly charged emotions and sometimes irrational behavior have figured into the search for the ivory-billed woodpecker. Informants for this book, for example, have described physical threats, lawsuits, and accusations of fraud against them after they have claimed to have seen the bird. Others said they would never go public even if they had seen the bird because their lives would never be the same. One landowner told me he would sooner shoot the bird than report its existence and have his land taken over by the federal government. Although federal authorities have never threatened to seize anyone's land, the fear of government intrusion surfaced often in my conversations with local landowners and outdoorspeople. Other people have taken unorthodox or extreme measures to find the ivory-bill, such as an animal psychic who claimed to have determined the exact location and number of ivory-bills around the time of the 2002 search in Pearl River (Gallagher 2005). Another Louisianan I interviewed offered a $10,000 reward for a photograph of the bird. It became obvious during the course of my research for this book that I was not writing about an ordinary bird.

Some local people who live in or near bottomland forests have claimed for years that ivory-billed woodpeckers survive in the Deep South. According to ornithologist James "Van" Remsen at LSU, his office had received reports of ivory-bills long before the Arkansas sightings. There have been other cases—the wolf (*Canis lupus*) in Maine and the jaguar (*Panthera onca*) in New Mexico—in which local people reported the reappearances of rare species long before wildlife experts and academics accepted their presence. Both of those species were reported by hunters. One possible explanation for the silence from wildlife professionals and nature advocacy groups about any remaining ivory-bills is that these people have kept the existence and location of the birds secret for fear that the birds and their habitat would be overrun by birders and the media. The fact that people involved in both the Arkansas and subsequent Louisiana searches were sworn to secrecy until authorities deemed the timing right to break the news would seem to lend support to this possibility. I have spoken with a couple of expert ornithologists and naturalists who said, off the record, that they are not sure they would even report seeing an ivory-bill today because it probably would not be in the best interest of the woodpecker to be bothered by hundreds of eager birders and media types. This admission might surprise the many people with a deep desire to see the bird, but it demonstrates the balancing act forced on ornithologists studying the ivory-bill, who must weigh their professional responsibility to report the species with their desire to protect it. Ornithologists have also spoken to me of their fear that their professional reputations would be questioned if they reported an ivory-bill sighting. Given the stakes involved—for both the bird and the professional credibility of academics involved in the controversy—it is not out of the question that credible sightings have occurred and gone unreported. The Arkansas sightings have demonstrated that the ivory-bill is more than a mythical ghost that lives only in the heads of imaginative birders, hunters, and fishermen.

Given the reluctance of some in the professional conservation fields to report their sightings, we must ask whether over the years we have set the bar too high in terms of what constitutes a credible sighting and missed some excellent opportunities to find the bird prior to the Arkansas search. This issue was raised by more than one of the experts I interviewed for this book. The contention that has attended the ivory-bill's rediscovery—with prominent figures in the ornithological community seeming intent on disputing claims of the bird's sighting despite the impeccable credentials of the members of the Arkansas team—has made scientists wary of entering the fray.

An example of the kind of controversy that may inhibit scientists' contributions to the search for the ivory-bill can be seen in the debate surrounding the article published in *Science* by author and birder David Sibley. A prominent skeptic, Sibley challenged the findings of the search team that had been presented in an earlier article in the same journal and claimed the bird in the film was a pileated woodpecker (Sibley et al. 2006). In response to my question about Sibley's criticism, an individual close to the ivory-bill search team responded: "We took his criticism seriously, but ironically, his pileated woodpecker in-flight illustration [in his field guide] is the least accurate I have ever seen in terms of ratio of white:black on the underwing. Anyone with a ruler can figure this out from his illustration." So even though Sibley is a respected artist and birder, the distortions in his drawing of a pileated woodpecker make his claim that the search team mistakenly identified a pileated woodpecker ring hollow among some of the search team. There are also questions about the accuracy of Sibley's interpretation of the Arkansas team's video. A search team member told me: "A major problem is that [Sibley's] hypothesis of showing that much white in the wing on both downstroke and upstroke is unlike any known model of avian flight. In fact, just watch the next bird you see flying away from you and ask yourself how much of the underwing you can see on the downstroke." But even though Sibley's interpretation can be refuted, the fact that the ivory-bill controversy seems to provoke this sort of high-profile public challenge may discourage some scientists from participation in the debate.

Jerome Jackson, one of the top woodpecker experts in North America, is another prominent skeptic of the Arkansas reports and film clip. In a 2006 article published in the *Auk*, Jackson lays out his arguments against the film having captured an ivory-bill, and against the ivory-bill being found in Arkansas specifically. Jackson's criticisms included that the search team had gotten the ivory-bill call and its drumming mixed up with that of the white-breasted nuthatch (*Sitta carolinensis*), and that the rediscovery announcement was timed to ensure maximum publicity for Tim Gallagher's book *Grail Bird: Hot on the Trail of the Ivory-Billed Woodpecker*, which is based on the Arkansas search and rediscovery. The scientific reputations of the leaders of the search team make it highly unlikely that they would fabricate their sightings or confuse them with other more common birds. In a follow-up article in the *Auk*, the Arkansas search team, led by John Fitzpatrick, disputes Jackson's criticisms point by point (Fitzpatrick et al. 2006).

Today there are two camps in the academic debate about the ivory-bill, both with positions firmly entrenched. In one group, there are ornithologists and birders, such as Jerome Jackson and David Sibley, who dispute the Arkansas sightings and the evidence presented in two leading journals, *Science* and the *Auk*. On the other side, there is the Cornell-led search team and the members of various groups associated with it who have no doubt about what they saw and the evidence they presented. Similarly, in the birding world, two distinct camps exist, the believers and nonbelievers.

The emotion, skepticism, and downright ill-feelings that exist in some circles between these camps have even led some very credible individuals who have reported ivory-bill sightings over the past thirty years to change their stories and to question their own sightings. Some seemed to try to talk themselves out of what they originally thought they saw. One wildlife official told me her sightings became a joke among her coworkers, and she was initially reluctant to share them with me. David Kulivan's story—related to me by Vernon Wright, one of Kulivan's professors—offers another example of the negative attention that can result from reporting an ivory-bill sighting. In 1999, Kulivan, a forestry student at Louisiana State University, claimed he saw a pair of ivory-bills while turkey hunting in the Pearl River Wildlife Management Area in Louisiana. Kulivan initially was reluctant to report what he saw for fear of the scrutiny and unwanted attention it might incur. Considering some of the response that his reported sighting elicited, his fear may have been well founded. Few birders or ornithologists stated in subsequent articles that they believed him. Some birders—people who had never met Kulivan—went so far as to claim in various birding chat rooms and list serves that Kulivan was a pathological liar who was probably hallucinating at the time. Perhaps now that well-known ornithologists have reported seeing the bird, reports like Kulivan's will meet with a different response.

The experiences of Professor George Lowery, a renowned ornithologist at LSU, offer an earlier example of the intensity with which skeptics have sometimes attacked ivory-bill reports. Lowery received photos of an ivory-bill that were taken in early 1971 by Fielding Lewis, an avid outdoorsman who lives in Franklin, Louisiana. Lowery believed the photos to be authentic. Because Lewis didn't want outsiders, especially the federal government, to find out where the birds still existed, Lowery refused to reveal the exact location where the photos had been taken or the identity of the photographer. While many ornithologists and birders believed the photos were authentic, others ridiculed them and questioned Lowery's intentions. According to some former students and colleagues, the accusations of fraud remained a source of tension for Lowery until his death in 1978. According to those who knew him, Lowery was a true scholar, making the accusations of fraud especially hurtful. The authenticity of the 1971 photos is still debated among birders and ornithologists.

Even though I believe there are a few small groups of ivory-bills in the bottomland forests, I am not particularly surprised that only one bird has been filmed and recorded. It is easy for skeptics to demand photographs, yet anyone who has spent time in bottomland forests in the South realizes that photographing an ivory-bill in this environment is far easier said than done. During most of the year, it is difficult to see more than ten feet into the forest from a road or bayou. And when one is paddling on a bayou or a cypress swamp, it would be next to impossible to get a photograph of a bird that flies over or through the timber. It is important to understand, too, that ivory-bills fly fast and straight, similar to a pintail duck. And tall trees line the banks and roadways in many prime ivory-bill habitats, giving birders mere glimpses of flying birds. Unless one were either to find a nest or roost hole where birds returned regularly or to sneak up on a feeding individual, a photograph would be hard to come by. Even the footage of the ivory-bill in Arkansas—shot knowing an ivory-bill was in the area— is distorted, offering testimony to the difficulty of producing a good photo. Compounding these challenges is the fact that the findings of the Arkansas search suggest that the ivory-bill is extremely shy. And yet somehow even the difficulty of photographing an ivory-bill, hidden away in its formidable forest home, contributes to the mystery and allure surrounding the bird.

The Power of Extinction and Rediscovery

Some 170 years after Audubon wrote about the ivory-bill, the present-day media, birders, and ornithologists continue to be enamored, some obsessed, with the bottomland ghost. An Internet search for the term "ivory-billed woodpecker" provides links to thousands of Web sites. These pages include the identification tips offered by the State of Louisiana to help the public in their search for the ivory-bill; personal Web pages dedicated to the current search or to testimonials about the bird;

and birder chat rooms where rumors of ivory-bill sightings are shared.

Perhaps this attention is not surprising given the emotions attendant on the idea of extinction—especially the extinction of "charismatic" animals such as large birds and mammals—and the finality it represents. The fates of the passenger pigeon (*Ectopistes migratorius*) and the Carolina parakeet (*Conuropsis carolinensis*) have taught birders and ornithologists that common species can be wiped out rapidly by modern firearms and destructive activities such as clear-cut logging. And it is generally thought that the ivory-billed woodpecker was never as common as these species. Given that the ivory-bill's territorial demands are expansive, it has been more vulnerable than either the Carolina parakeet or the passenger pigeon ever were (Tanner 1942a).

It is disturbing to realize that, in this age of supposed environmental awareness, the ivory-bill was almost permanently reduced to being another museum exhibit like the passenger pigeon. Today the ivory-bill remains critically endangered, although many former ivory-bill habitats are protected as state parks, national parks, wildlife-management areas, national forests, and other conservation areas. In fact, though it is a fraction of what existed two hundred years ago, there is more prime ivory-bill habitat located in protected areas today than at any time in the past one hundred years (Gardiner and Oliver 2004; Jackson 2004; Nature Conservancy 2006). The ivory-bill's prospects might be brighter today if targeted habitat-conservation measures had been implemented at the beginning of the twentieth century.

In terms of habitat requirements, the ivory-bill needs mature bottomland forest, with enough dead trees to house the beetle larvae that infest dead and dying trees (Tanner 1942a; Jackson 2004). Managing a forest to provide enough sick or dead trees for a small population of ivory-bills would be feasible. It was noted in the past that ivory-bills would invade a forest after it burned or after large trees were deadened by early settlers in an attempt to make land clearance easier. While analyses of the ivory-bill's stomach contents by past researchers have shown beetle larvae to be its most important food source, the bird also eats a variety of other resources such as muscadine grapes, persimmons, hackberries, acorns, and the berries of magnolia, poison ivy, and tupelo gum trees (Tanner 1942a). These resources are common in today's bottomland forests in the South. According to Van Remsen at LSU, the age of a forest, or at least of many of the trees in the forest, is perhaps the most important overall factor regarding the presence of ivory-bills. For example, in James Tanner's study of the ivory-bill in the 1930s, he noted that it feeds primarily on large trees, with 87 percent of the birds in his study feeding on trees that were more than one foot in diameter (Tanner 1942a). The recent sightings in Arkansas and Louisiana testify to the resilience of nature. Perhaps enough wild areas still exist to provide sufficient habitat for the ivory-bill to survive.

The rediscovery of a species thought to be lost is a compelling notion in an age of postdiscovery in the United States (Cokinos 2000; Weidensaul 2002). How many "extinct" animals have been found again? There are seemingly few places on the planet, let alone in the United States, that have not been mapped and described or at least visited by scientists, tourists, and developers.

Might we dare hope that in addition to the Arkansas and Louisiana ivory-bill reports, the earlier Pearl River report was authentic too? Is it possible that just north of New Orleans, near Interstate 10 no less, other hidden biological treasures still exist? Are there wild enough landscapes in our own country, our own backyards, to house species such as the ivory-billed woodpecker? I submit that the answer is yes. Nature is once again proving to be more mysterious than we originally thought. We may have arrogantly underestimated nature's ability to adapt to changing circumstances, in this case meaning the ivory-bill's ability to survive in largely secondary, degraded forest. Perhaps a few ivory-bills, or even a few small populations, have successfully adapted to a rapidly changing and degraded environment. After all, most of our present-day conclusions about ivory-bill ecology and behavior are based only on a single study conducted in the late 1930s by James Tanner (Tanner 1942a). It seems there are a few birds still out there, and that they are better adapted to the modern world and its fragmented environment than we realized.

2

BOTTOMLAND FORESTS

A Southern "Paradise Lost"?

The decline of the ivory-billed woodpecker is symbolic of man's degradation of the wild southern environment, especially its bottomland forests. Found between land and water, the bottomland forests are one of the most diverse ecosystems in North America (Bedinger 1981; Martin, Boyce, and Echternacht 1993; Twedt and Loesch 1999). Estimated to have covered more than 52 million acres of the Southeast before the arrival of the Europeans—with 24 million acres in the Lower Mississippi alluvial valley alone—these forests once teemed with wildlife (figure 3). In 2005, less than 5 million acres remained (map 2) (Nature Conservancy 2006). Thus, historically, the ivory-bill was found in much of the South, especially around rivers in the bottomland forest zones.

These swamps have captured the imagination of many southern writers, travelers, and residents. Some early accounts describe the forests' wealth, especially their wildlife, while others saw the bottomlands forests as menacing environs not fit for human inhabitation. Audubon (1999:270), providing one of the more colorful portraits of bottomland forests, described the specific habitat of the ivory-bill in his *Ornithological Biography* (1831–39):

> I wish, kind reader, it were in my power to present to your mind's eye the favourite resort of the Ivory-billed Woodpecker. Would that I could describe the extent of those deep morasses, overshadowed by millions of gigantic dark cypress, spreading their sturdy moss-covered branches, as if to admonish intruding man to pause and reflect on the many difficulties which he must encounter, should he persist in venturing farther into their almost inaccessible recesses, extending for miles before him, where he should be interrupted by huge projecting branches, here and there the mossy trunk of fallen and decaying tree, and thousands of creeping and twining plants of numberless species!

Indeed, the bottomland forests and their giant cypress sentries provided habitat for a myriad of large and small animal species (figure 4). Old-growth cypress, tupelo gum, and sweet gum (*Liquidambar styraciflua*) trees often averaged between four and six feet in diameter. While some of these giants remain in protected or inaccessible swamps, most of the old-growth bottomland forests were cut by the early twentieth century to feed America's insatiable appetite for wood and agricultural products. Today much of this land has been rendered into relatively isolated islands of forest, usually surrounded by soybean and cotton fields and commercial catfish ponds (Saikku 1999) (figure 5). And yet, despite the selective logging that has occurred during the past two centuries and the resulting geographical isolation of tracts of bottomland forest (Dennis 1988; Carter Ewel and Odum 2001; Nature Conservancy 1992), ivory-bill habitat still remains.

Given the swamps' abundance of mosquitoes, snakes, and alligators, few people would be likely to consider bottomland forests a paradise, but the great biodiversity they house makes them a biological treasure. The Tensas River National Wildlife Refuge in northeastern Louisiana, for example, is home to more than four hundred species of mammals, birds, reptiles, amphibians, and fish (Barrow 1990; Nature Conservancy 1992). President Teddy Roosevelt, while bear hunting in the Tensas River area in 1907, commented on the great trees that he saw in what were once the most impressive forests in the eastern United States: "In stature, in towering majesty, they are unsurpassed by any trees in our eastern forests; lordlier kings of the green-leaved world are not to be found until we reach the sequoias and redwoods of the Sierras" (Schullery 1986:87). But even the admiration of a president was not enough to save the Tensas forests.

While bottomland forests are biologically important, these ecosystems have not been treated as such. Over roughly the past two hundred years in the South, these forests have been cut, drained, plowed, and confined with levees—all in an effort to gain access to their timber, wildlife, and rich soil. With the fate of the ivory-bill tied so directly to the condition of the bottomland forests, the bird may be thought of as an indicator species for this unusual habitat. An examination of the history of human use of the bottomland forest, therefore, may do more than illuminate the reasons for the ivory-bill's decline. Understanding the way our use of these forests has pushed the bird to the brink of extinction may help us discover forest-management techniques that could be key to bringing the bird back.

Bottomland forests in the Deep South, especially along the Mississippi River, were extensively cleared beginning in the early 1800s to make way for the great plantations whose livelihoods depended on cotton, sugarcane, and slave labor (McWilliams and Rosson 1990; Saikku 1996, 2005; Hamel and Buckner 1998), and the logging of the once-great expanses of bottomland forests is the factor that has likely had the greatest impact on the ivory-bill (Tanner 1942a; Christy 1943; Hamel and Buckner 1998; Saikku 2001; Hoose 2004). Forest clearance started well before the Civil War, but the period between the end of that war and World War II witnessed the most intensive destruction of the bottomland forests. Certain political and ecological factors in the South were responsible for this. First, the federal government sold vast tracts of forest lands to private owners, mainly northern logging companies (Sternitzke 1976; M. Williams 1982; Cowdrey 1995; Hamel and Buckner 1998). For example, the Singer Sewing Machine Company of Chicago owned the area (formerly referred to as the "Singer Tract") that is now the Tensas River National Wildlife Refuge (figure 6). Because most of the northern and eastern forests already had been cleared, northern companies turned their attention to the forests of the Deep South (Saikku 2001; Hoose 2004).

Lack of old-growth forests in the East, coupled with technological innovations in logging, allowed greater and more efficient access to forest resources—even previously inaccessible swamp forests. The overhead skidder was developed in the 1880s. Overhead or cableway skidders hauled logs out above the forest floor by connecting cables between two points above the ground. Since the logs no longer had to be hauled overland by mules and oxen, this device allowed logs to be taken out of wetland areas above the mud and stumps more quickly. The invention of the pullboat logging system at the beginning of the twentieth century also contributed to the invasion of swamp forests by loggers. And finally, the introduction of the chain saw to the logging industry in the late 1940s further accelerated the destruction of the old-growth bottomland forests, of which only shrinking patches remained by that time (Bean 1986; Saikku 2001).

In addition to new logging technologies that made cutting more efficient and profitable, bottomland forests were made more accessible by large-scale efforts to drain swamps and build levees along the great rivers in the South. Obviously, once previously flooded forests were drained and relatively dry, their timber became much more accessible to loggers. For example, following the great flood in 1927 along the Mississippi River, the federal government set out to restrict the river's ability to overflow its banks through the construction of a massive levee system (Hudson 1998). Although the levee system had begun before the great flood, the damage during the 1927 event spurred on more extensive levee construction. Because seasonal flooding had been restricted, adjacent forests not only became accessible to loggers but also were changed ecologically by the alteration of their hydrology (Bedinger 1981; Nelson and Sparks 1998; Conner and Brody 1989; Saikku 2005:252–53). Species such as willow (*Salix nigra*) have invaded many altered areas, forming almost single-species stands. Similarly, levees built around the Atchafalaya Basin following the 1927 flood confined water within a designated area, not a natural one. As a result, land that was seasonally inundated on the edges of the basin became high and dry, making it accessible to loggers, who were followed by farmers.

The Atchafalaya Basin's hydrology also changed as a result of the confinement. Over the past seventy years, countless battles have been fought between commercial fishermen, sportsmen, and the Corps of Engineers over water levels within the basin (Reuss 2004). Water levels appropriate for navigation are not always good for fishing or wildlife. The new, higher water levels that resulted from confining the floodwaters were incompatible with permanent human settlements (Comeaux 1972; Abbey 1979; Reuss 2004). Today, the only permanent town that remains in the basin is Butte la Rose.

The logging booms driven by the need for wood products during both of the world wars was another historical factor leading to the degradation of bottomland forests (Hoose 2004). Thus, remaining stands of primary forest were targeted by logging interests, and conservation and endangered species took a back seat to the needs of a nation at war. With a national conservation culture in its infancy during the early and middle twentieth century (and even nonexistent in many regions), there was little public resistance to the destruction of these forests. This was clearly evident during the destruction of the former Singer Tract, the last remaining known territory of the ivory-bill and one of the last (if not the last) large stand of old-growth bottomland forest left in the South. While the plight of the ivory-bills in the Singer Tract was not ignored—national conservation leaders had called for the logging to be halted—company officials ignored their pleas, citing the nation's war needs, and the forest was largely cut over (Hoose 2004; Jackson 2004; Gallagher 2005). According to the Nature Conservancy (1992), no other wetland system in North America has undergone the level of destruction that has occurred in the bottomland forests.

Even though much of the bottomland forests was destroyed by the mid-twentieth century, activities such as logging, dredging, draining for irrigation, and conversion to soybean fields continue to this day to eat away at some remaining forests, especially on private lands. The Tensas River Basin in northeast Louisiana, home to James Tanner's famous 1930s study of ivory-bill ecology, lost 182,500 acres of forested wetlands between 1979 and 1987—mainly due to the expansion of soybean production (Sternitzke 1976; Nature Conservancy 1992). While much of the forest that was felled during this time was second growth, it nonetheless was a regenerating bottomland ecosystem. And in the Grand River area in the northeastern section of the Atchafalaya Basin (figure 7), logging has taken place over the past two decades and continues today in this area that has produced many solid reports of the ivory-bill, one as recently as the fall of 2005. Even many protected areas such as state management areas and national wildlife refuges in Louisiana (Tensas, Pearl River) are still subject to occasional selective logging.

Today, bottomland forests cover a fraction of their pre–European Contact range (Hamel and Buckner 1998; Hamilton, Barrow, and Ouchley 2005). For example, in the larger Mississippi River alluvial valley in eastern Arkansas, the area of the 2004–5 sightings, only 15 percent of the state's original Mississippi bottomland forests remain. Similarly, the state of Mississippi has only 15 percent of its original wetland forest remaining; Missouri has 14 percent; followed by Tennessee and Illinois with 9 percent, and Kentucky with 4 percent (Nature Conservancy 1992). There were similar losses of this prime ivory-bill habitat in other southern states such as Florida, Georgia, and South Carolina. The shrinking forest patches obviously had negative consequences on the ivory-bill and other species, like bears, that need large tracts of territory. And within the forested wetlands of the Lower Mississippi River, 31 percent of all of the remaining forested wetlands survive in a single location: the Atchafalaya Basin in southern Louisiana. Given the size and isolation of the basin, it is hardly surprising that so many solid reports of ivory-bills have come from this area. In the Atchafalaya Basin, where mature secondary forest has reemerged and some old trees were left standing, reminders of a glorious past are everywhere in the form of old, weathered, and rotten cypress stumps slowly giving in to the elements.

With so little truly old-growth bottomland forest remaining, some experts even argue that no virgin bottomland forest exists at all (Fitzpatrick, personal communication). While the second-growth forest is impressive in some areas, it is not pristine. The huge old trees scattered among younger second-growth forest give some areas the look and feel of a virgin forest. These old trees were left by loggers, however, because they were either hollow or crooked and therefore had little timber value. Cat Island National Wildlife Refuge near St. Francisville, Louisiana, is one such area. Huge cypress trees are scattered throughout, the remnants of a primeval forest felled long ago. The largest living cypress tree, in fact, is protected on Cat Island. When one wanders through the huge trees in places like Cat Island, it is hard to believe that the forest is not old-growth. One almost expects to hear the ivory-bill's call, as Audubon had many years before. But most of that forest was logged repeatedly over the past century. The rapid growth of tupelos, coupled with the cypress giants, easily misleads the casual observer. While these areas are not the pristine, old-growth forest that constituted the ivory-bill's original habitat, the rediscovery in Arkansas demonstrates that if some mature trees are left standing, the bird may be able to hang on.

The upland pine forests adjacent to the bottomlands have been equally degraded. To the east of the Delta, along Interstate 55, are huge stands of pine trees. These trees that make for such monotonous driving, hour after hour, on the interstate are single-species, even-age pine plantations. And while these upland forests were probably not as important to ivory-bills as the adjacent bottomland forests, the presence of these pine plantations is nonetheless one more indicator of the modern-day destruction of the southern environment.

It is important, however, not to glamorize past environments as being untouched by humans. Not all of this bottomland forest was undisturbed prior to European settlement. Native Americans did indeed influence the age, structure, and composition of past forests (Stanturf et al. 2001). Native American agricultural practices resulted in a patchwork of forests in different stages of regrowth (Saikku 2001, 2005). But it should also be noted that Native American land-use practices were conducted on a smaller scale than the land clearing pursued by Euro-American settlers. Stone axes and fire cannot compete in terms of their scale and damage with chain saws, bulldozers, and levees. The Native Americans' small-scale burning of forests—to clear them for crops and to create better grazing for game animals—led to some deforestation, but that may have actually provided some habitat advantages for the ivory-bill. Old trees damaged or killed by fires and left standing would have provided both nesting and food for ivory-bills.

It is also important to acknowledge the factors other than the logging of the southern bottomland forests that have pushed the ivory-bill toward extinction. In addition to the direct destruction of their habitat, ivory-bill numbers were also reduced by human hunting and specimen collecting. Hunting of the ivory-bill has been traced to Native Americans who sought the bird's skins, bills, and skulls. Ivory-bill remains have been found far outside the bird's natural range as far west as Colorado and as far north as Canada, indicating that the ivory-bill may have been a coveted trade item among Native Americans who lived far outside its natural range and who had never seen a live bird (Audubon 1834; A. M. Bailey 1939; Parmalee 1958). The early American ornithologist and natural historian Mark Catesby (1985) commented that the princes and warriors of Canadian Indians made coronets out of ivory-bill remains. He also mentions that northern tribes traded two or three buckskins to their southern counterparts for each bill. The early European explorers and settlers appeared to have shared this interest in ivory-bill remains. While traveling down the Mississippi River, Audubon commented that the head of the male bird, in particular, was a popular curio among river folk, who often would "pay a quarter of a dollar for two or three heads of this woodpecker" (Audubon 1999:272). While it is unlikely that Native Americans or even early European Americans would have driven the ivory-bill to extinction—habitat destruction was the driving force in that process—they did no doubt reduce numbers, which could have led to local extinctions. When the ivory-bill was still fairly common, for example, Native Americans could have easily identified nest or roost trees from which birds could be taken without much difficulty. The impact of Native Americans on the species likely intensified after firearms were introduced in the seventeenth century. And even cultural groups that did not especially value the ivory-bill began hunting them once they become aware that other Natives and European settlers would either purchase the remains for cash or trade other valuable items for them (Catesby 1985).

While early hunting reduced and in some cases decimated ivory-bill numbers in the South, hunting didn't push the bird to the brink of extinction until the late nineteenth and very early twentieth centuries. This was not a case of populations being decimated because thousands of birds were shot; by that time, thousands of ivory-bills no longer remained. In isolated and declining populations of the birds, however, removing just a few individuals could decimate a population. There is no dispute that radical environmental alteration took place in the territory inhabited by the ivory-billed woodpecker, and certainly this habitat destruction made the ivory-bill more vulnerable to other forces such as hunting. And, while hunters may not have specifically sought out the ivory-bill, when times were hard, any large bird was fair game.

The ivory-bill probably served as food into the nineteenth and early twentieth century, when rural residents depended more on wild game than do today's recreational hunters (Eastman 1958). For example, Wayne (1895) reports that in Florida, ivory-bills were considered quite desirable as food by the local residents. In targeting the birds, hunters participated in a long tradition in the Deep South and elsewhere in our country of exploiting a vast array of wild game as food. This tradition was especially strong in southern Louisiana. The Cajun and rural African American cultures in southern Louisiana depended more heavily than the larger society on natural resources. The Cajun name for the ivory-bill is *pique-bois grande.* And if there is a Cajun name for a bird, there is probably a recipe, or at least a gumbo. Both poverty and proximity to wild areas led rural folk to depend on game animals because cash and retail stores were in short supply. Many older informants talk about the days when blackbirds and various large birds, including woodpeckers (also called "forest chickens"), were sought out in the nearby forests and swamps, especially in the 1930s. While wild game consumption is not necessarily as common or driven as much by poverty as it once was, rural Louisianans probably still consume more wild game and nongame animals today than is usual in most other states. For example, in south Louisiana, the yellow-crowned night heron (*Nyctanassa violacea*) is eaten by many, as are crawfish, turtles, frogs, and garfish. The state motto, "Sportsman's Paradise," thus applies not only to typical game animals such as ducks and deer but also to creatures that people in other regions of the United States might find less palatable.

Hunting in the South has a dark side as well. Numerous historical studies describe excessive takes of game species such as deer, waterfowl, and small game in the South. In his *Subduing Satan: Religion, Recreation and Manhood in the Rural South* (1990), Ted Ownby recounts the plunder of southern hunts from 1865 to 1920. Bear, deer, small game, waterfowl, and various birds were killed in huge numbers, without any regard to the conservation of future stocks. Ownby argues that these perverse hunts were partly a response to the increasing social control of men and their activities by women and evangelical religious figures. Hunting placed men beyond the controls of this conservative, feminine culture and allowed them to act as they pleased, even excessively—in terms of quantities of animals killed and the methods employed in the killings. Today, such excesses are rare, and ivory-bill conservation is compatible with hunting and fishing activities. Certainly healthy bottomland forests provide habitat for innumerable fish and game species. Also, because of the money spent on their sports, hunters and fishermen represent a powerful lobby interest in states such as Louisiana. If these groups were encouraged to throw their support behind the protection of ivory-bill habitat, the effort would stand a much greater chance of success.

Ivory-bill populations were also reduced by specimen and egg collecting. In fact, scientific hunting likely had a larger impact than game hunting (Tanner 1942a; Hoose 2004; Jackson 2004). It is ironic that the birding and scientific communities—the very individuals one would expect to protect the bird—sought more and more specimens as ivory-bill numbers dwindled in the early twentieth century. During this time, collecting and mounting rare species and their eggs was the fashionable way to show an appreciation for nature—especially birds. In fact, as the ivory-bill became rarer, naturalists began to compete for specimens, resulting in hundreds of birds being killed for private and museum collections (Jackson 2004).

Despite the presence of other factors driving the bird toward extinction, the direct destruction of their habitat through logging had the most significant impact on the remaining ivory-bill populations during the late nineteenth and early twentieth centuries. Lingering ivory-bill populations living around the South likely struggled to survive in the remaining patches of forest. And the isolated tracts of bottomland forest, in turn, would have made ivory-bills increasingly vulnerable to the secondary ecological impacts of logging, such as greater predation and genetic problems because of the small breeding pool. If few mature trees with nest holes survived in a particular location, it is likely that the remaining ivory-bills competed for nesting locations with species such as pileated woodpeckers (Cowdrey 1995; Saikku 2001).

Jerome Jackson, a noted woodpecker authority, has written about the potential conflict between pileated and ivory-billed woodpeckers in shrinking forests (Jackson 1989, 2004). A host of other birds and animals also might have competed with the ivory-bill for nest sites, including small owls, wood ducks, mammals, and even honeybees, who build hives in hollow trees. Ivory-bill nestlings and eggs were probably preyed upon by a similarly diverse array of animals. If forests were degraded and reduced in size, ivory-bill nests would likely have suffered "edge" effects due to predators such as raccoons, opossums, snakes, and other animals and birds that thrive in such environments. Studies of neotropical migrants have shown that edge predation has significantly impacted the breeding success of certain species (Andren and Angelstam 1988; Laurance and Yensen 1991). While the ivory-bill is much larger than a warbler, its nests would similarly be vulnerable to detection and predation. And the genetic isolation of not only ivory-bills but of all organisms that inhabit forest islands is another important impact of habitat alteration and destruction of bottomland forests. In areas where ivory-bills survive, one must question the genetic fitness of these small populations. Although some long-range movement of birds has almost definitely taken place, given the small numbers of ivory-bills, inbreeding depression, which reduces fitness, is an issue that may threaten any long-term recovery of the ivory-bill.

Even with all these negative impacts on its habitat, however, it appears that enough bottomland forest was left standing so that some ivory-bills found refuge—in Arkansas, Florida, and in Louisiana. But given the simple fact that this environment underwent wholesale destruction and the bird's numbers are likely very small, there is no doubt that the ivory-bill's existence is tenuous. And even with the euphoria over the Arkansas sighting, it is critical that we remember that this species is in grave trouble. If there is a breeding population in Arkansas and elsewhere, it is imperative that state and federal authorities more closely monitor land-use practices in areas that hold potential for ivory-bills such as Pearl River Wildlife Management Area and the Atchafalaya Basin. While selective logging does not necessarily spell doom for ivory-bills, climax forest (or at least dead and dying timber within a forest) must be allowed to develop.

Thus one of the most important issues that must be addressed regarding the southern bottomland forests that comprise the ivory-bill's habitat is the impact of current forestry-management practices on tree species composition and forest structure (Shoch 2005). Are current forestry practices appropriate for producing trees desired by the ivory-bill, such as old dead and dying sweet gums? Or are current practices focused on producing commercially desirable timber such as various oak species—species that might not provide the best ivory-bill habitat? Structural diversity—a trademark characteristic of old-growth forests—is lacking in today's reforested landscapes, where even-age trees dominate forests (Hamilton et al. 2005).

Chuck Hunter of the U.S. Fish and Wildlife Service broached this topic in a 2005 working paper discussing future forest-management practices. Although consideration of nongame species and tree species' diversity has been given greater emphasis in the past twenty years, it remains unclear if we are growing forests suitable for ivory-bills. The rediscovery of the ivory-bill may be a watershed moment in forest-management practices on federal and state lands in the South. It may lead to federal and state resource managers reexamining current forestry paradigms and to important discussions between these agencies and private landowners concerning the management of potential ivory-bill habitat.

Large-scale reforestation efforts in bottomland areas have already been initiated by federal, state, and private groups and programs such as the federal Conservation Reserve Program and Wetland Reserve Program (Kennedy 1990; Gardiner and Oliver 2004). In most instances, these efforts do not appear to be re-creating pre–European Contact forests, although determining the exact species composition of truly pristine forests may be impossible (Stanturf et al. 2001). We do at least know that the species composition of bottomland forests of one hundred years ago is different from what is developing today (Savage, Pritchett, and Depoe 1989; Ouchley et al. 2000; Haynes 2004). In today's reforestation efforts, over 80 percent of trees replanted are oaks and pecan (Twedt et al. 2002). Another study found that oaks made up over 90 percent of the trees replanted in the national wildlife refuges in Arkansas, Louisiana, and Mississippi (Ouchley et al. 2000). Oaks and pecan were certainly present in old-growth forest and provide many environmental services, but these services—producing deer feed and timber—appear to be directed more at human use than at creating a diverse (structurally and biologically) bottomland forest community (J. A. Allen 1990, 1992, 1997; Newling 1990; Ouchley et al. 2000; Hamilton et al. 2005). It was noted in the U.S. Army Corps of Engineers' study of bottomland forest restoration that "it is unlikely that restored areas reflect the structure and function of the original forests" (Guilfoyle 2001:11).

When one visits Tensas River National Wildlife Refuge today, it is hard to believe that the isolated (albeit large) forest patches that sit among soybean fields and human poverty so prevalent in the Delta were once home to a thriving ivory-bill population. Some of this clearing was part of various forest-management strategies, such as selectively removing certain species or age categories to enhance wildlife habitat. But many of the professionals that I interviewed for this book questioned whether these management strategies will ensure the creation of habitat for species that need old forest—such as the ivory-bill, which depends on dead and dying timber—or that depend on structurally complex forests (Hamilton et al. 2005). While these experts commended the bottomland reforestation efforts because former croplands are returning to forest, many stated that it is imperative that future forest-management plans create more old-growth and more ecologically and structurally complex forests (Hamilton et al. 2005; Stangel 2005).

And, while the Arkansas sighting proves that the ivory-bill can survive in partially degraded and secondary forest habitat, we should not view that as an excuse to continue our current land-use practices in the maturing stands of bottomland forest in certain areas of the South. It would be unreasonable to demand a moratorium on all logging activities on private lands, but every logging proposal for public lands should be reviewed in areas where ivory-bills might still exist.

Given all of these direct and secondary human-driven impacts—logging, hunting, specimen collecting, potential genetic problems—we are fortunate that any ivory-bills survive today. It seems the ivory-bill is more resilient and flexible than previously believed. This flexibility should not give us comfort or create complacency in our efforts to protect the bird and its remaining habitat. Instead, the rediscovery should give us pause and make us appreciate the fact that we have been given a chance to ensure the ivory-bill has a future.

Fig. 1. Controversial ivory-bill photos taken by Fielding Lewis near Franklin, Louisiana, in 1971 and sent to George Lowery at LSU. Note the two different tree species and the slightly different angle at which the bird clings to the tree. The authenticity of these photos has been heatedly debated for more than three decades. Lowery and many others believed them to be authentic, yet critics claimed the birds were mounted specimens placed in the trees. Photo courtesy of Tommy Michot.

Fig. 2. Aerial photograph of bottomland forest near Patterson, Louisiana, where ivory-bills were spotted and heard in 2005. This area has produced many ivory-bill sightings over the past several decades. Photo courtesy of Tommy Michot.

Fig. 3. Old-growth bottomland forest with Wylie Barrow in Bayou Cocodrie National Wildlife Refuge. This refuge protects one of the last stands of old-growth bottomland forest in the Mississippi Delta region. Photo courtesy of Wylie Barrow.

Fig. 4. Flooded bottomland forest in the Tensas River National Wildlife Refuge. This is a difficult environment in which to search for an elusive and rare bird. Photo courtesy of Wylie Barrow.

Fig. 5. Mature second-growth bottomland forest in the Tensas River National Wildlife Refuge. Even though the forest may be only sixty years old, thanks to the climate and rich soils, it now resembles a grand old forest. Photo courtesy of Wylie Barrow.

Fig. 6. A male ivory-bill at a nest hole. This photo by James Tanner was taken in the former Singer Tract. Tanner's photos are the last widely accepted photos of the ivory-billed woodpecker. Tensas River National Wildlife Refuge, U.S. Fish and Wildlife Service, Ivory-Billed Woodpecker Records, Mss. 4171, Louisiana and Lower Mississippi Valley Collections, LSU Libraries, Baton Rouge, LA.

Fig. 7. Logging in the Grand River area in the eastern section of the Atchafalaya Basin, 1980s. This area has produced many reported ivory-bill sightings over the past few decades. Photo courtesy of James "Van" Remsen.

Fig. 8. Identification key for ivory-bill and similar species. Courtesy N. John Schmitt.

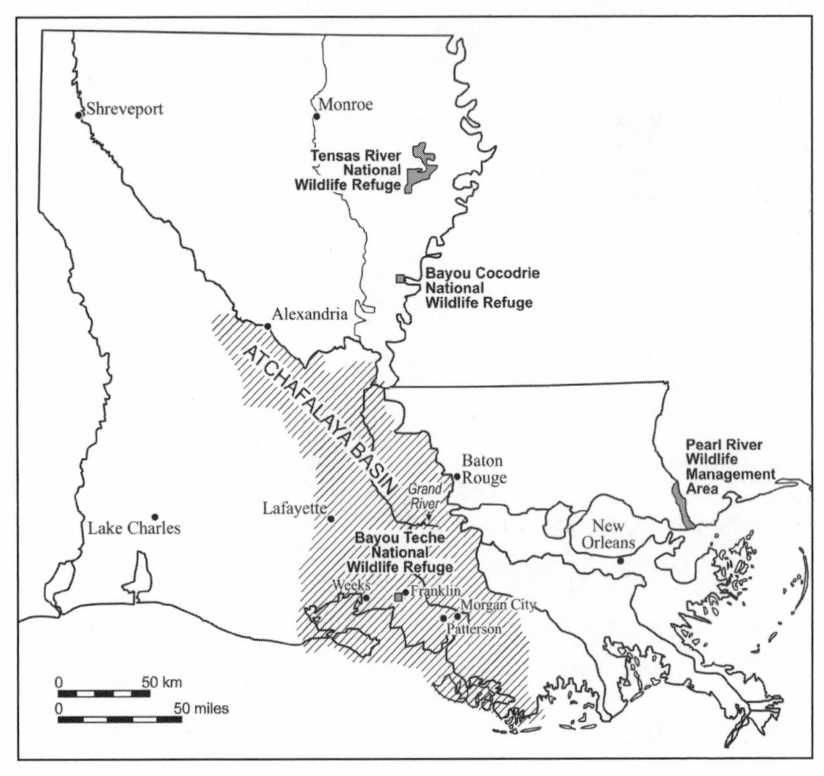

Map 1. Significant ivory-bill locations in Louisiana.

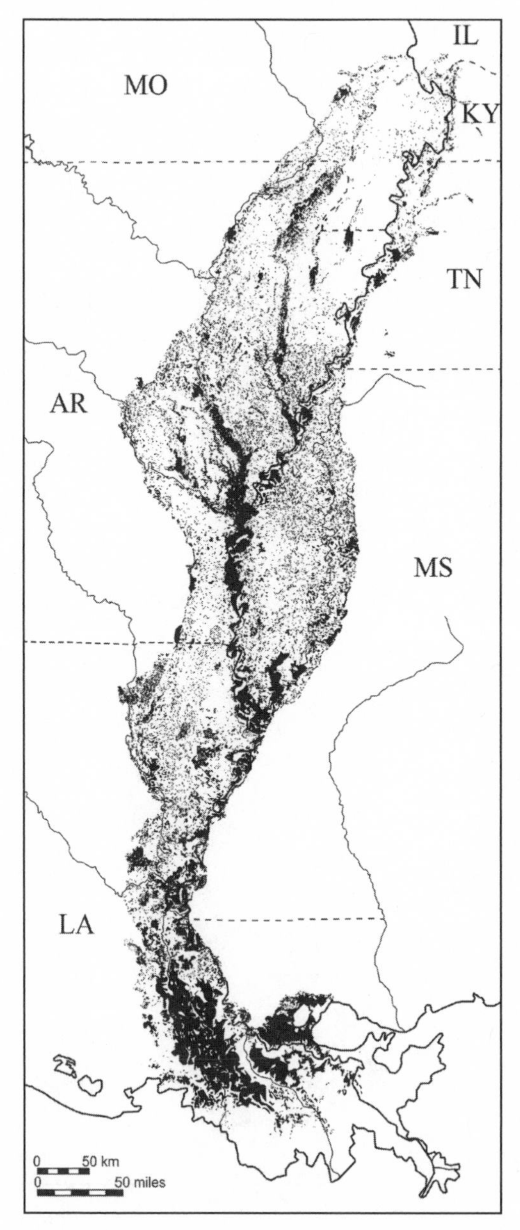

Map 2. Forest cover in the Lower Mississippi alluvial valley.

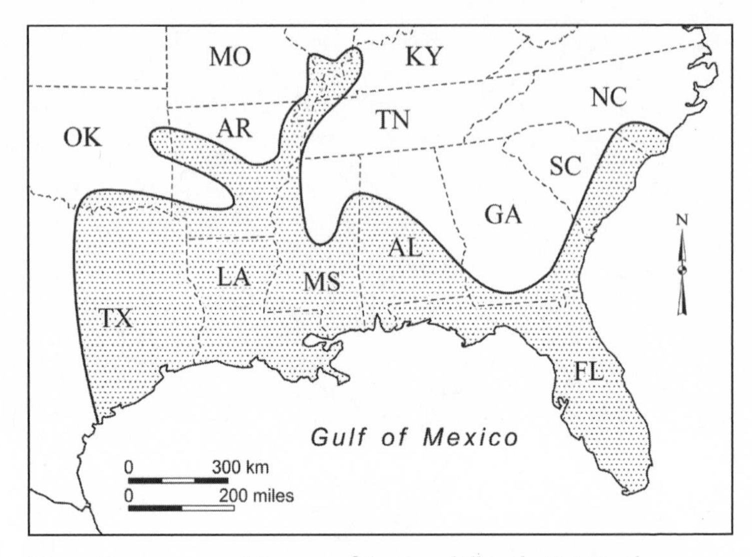

Map 3. Former geographic range of the ivory-bill in the U.S. South.

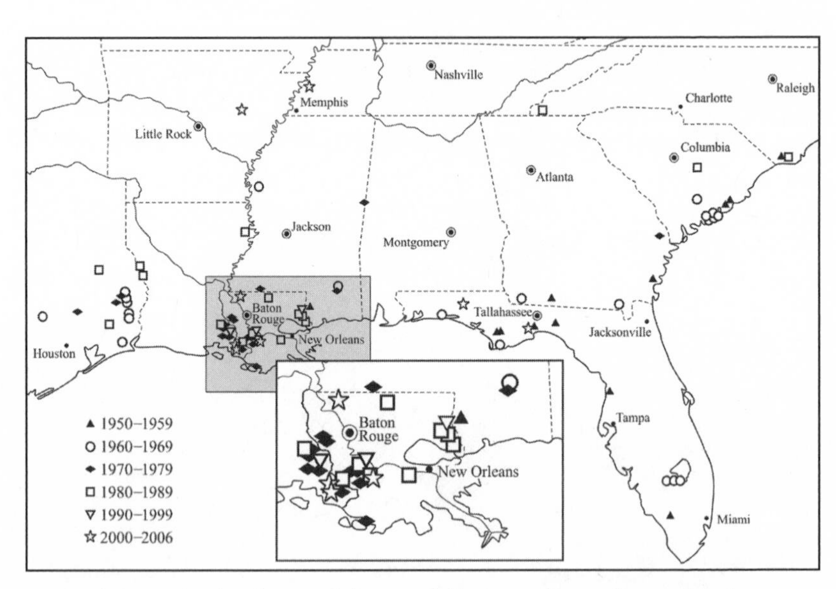

Map 4. Past reported sightings of the ivory-bill, mapped according to decades, 1950–2006.

3

THE EXPERTS

While researching this book, I sought out conservation personnel, resource managers, and ornithologists to provide expert opinions on the ivory-bill. What did they think of past reported sightings, the 2005 Arkansas news, and the prospects of finding the ivory-bill in Louisiana? The resulting discussions with experts were both exciting and discouraging.

The interviews were exciting when an ornithologist or professional conservationist said that he or she had seen an ivory-billed woodpecker. Such conversations made it easy to believe the bird exists—not only in Arkansas but also in places like the Atchafalaya Basin in Louisiana. Ornithologists and conservation officials have a great deal to lose professionally when claiming to see a bird that was thought to be extinct. This is borne out by the caution with which the Arkansas search team proceeded after their sightings of the ivory-billed woodpecker. They withheld information about their sightings until they had proof in the form of a video clip of the bird. They realized their colleagues would, at best, chuckle behind their backs and claim that they had caught what some Louisiana folk call "swamp fever" after spending too much time in the swamps.

On the other hand, when some very qualified experts threw their professional weight behind their opinion that the ivory-bill is gone for good and that the reported sightings are just unverifiable stories, the conversations were depressing. Even after the 2005 Arkansas reports, some professionals remain skeptical that the search team really saw an ivory-bill. And among those who believe that a few birds may still exist, some are especially doubtful about the long-term viability of a small ivory-bill population.

The following chapter consists of my interviews with various experts, mostly, but not exclusively, those with interest and experience in Louisiana. The interviews took place both before and after the Arkansas rediscovery news was published in *Science* on April 28, 2005. These interviews are significant for two main reasons: First, it is important to understand why, even before the Arkansas reports, some individuals felt the ivory-bill survived. The reasons these experts give for their optimism have important implications regarding other searches and the long-term recovery possibilities of the ivory-bill. In the interviews, I explore the environmental and other factors that contributed to this optimism, and whether they think the ivory-bill is likely to exist outside of Arkansas. And second, the skeptics raise significant questions about the present and future status of the bird. Finding one bird or a handful of birds does not guarantee the long-term survival of the species. If the number of remaining ivory-bills is critically small (which is likely), then the skeptics' views about its future need to be understood and addressed in any attempt to preserve the species.

The material contained in this chapter and the following one is based on interviews I conducted in person, on the telephone, and via e-mail between 2003 and 2006. The only secondary material is David Kulivan's note describing his 1999 ivory-bill sighting, which he gave to Professor Vernon Wright at LSU, who then shared the information with me.

Keith Ouchley: The Nature Conservancy

Keith Ouchley, the director of the Louisiana chapter of the Natural Conservancy, has many professional, personal, and familial ties to the ivory-billed woodpecker and its bottomland home. First, the Louisiana Nature Conservancy owns and manages some bottomland forests where the ivory-bill is known to have lived until at least the beginning of the twentieth century. Ouchley also wrote a dissertation focused on neotropical migrant use of the bottomland forest habitat in Louisiana, and his brother served as a warden at the Tensas River National Wildlife Refuge, the location where Tanner completed the ivory-bill study. More important, in 2005–6, during the time I was researching this book, Ouchley was overseeing the formal search effort for the ivory-bill near Patterson, Louisiana, and the formation of the State of Louisiana's Ivory-billed Woodpecker Task Force.

I first met Keith Ouchley for lunch in historic St. Francisville, Louisiana, in July 2003. Ouchley is a serious environmentalist, who, while fascinated with the ivory-bill, was not caught up in the emotion surrounding reported sightings prior to the Arkansas announcement. He said he dearly wanted to believe the reports of the local sightings, but, given that he has extensively studied the very forests that once housed the ivory-bill, his thoughts on the bird's continued existence were guided by science and not dreams. Even though he knew personally some of the individuals who claimed to have seen an ivory-bill, he would not let himself get too excited about reported sightings without proof.

Although Ouchley said that his entrenched skepticism was hard to shake, the release of the Arkansas report made him a believer that the ivory-bill exists. But even while he is now confident that there are ivory-bills in Arkansas and Louisiana, he is still guided by science and not emotion regarding the bird's future. A self-described realist, Ouchley is concerned by the fact that the ivory-bill's numbers are extremely low.

It was appropriate that Ouchley and I met in St. Francisville to talk about the ivory-bill because the town is steeped in history relating to Audubon. It was in St. Francisville that Audubon wrote about the ivory-bill and its bottomland home. The plantation at which Audubon stayed has been preserved, and his name graces a local state park, a nightclub, and even a liquor store. St. Francisville has promoted its connection with Audubon and the ivory-bill in an effort to attract tourists. The ivory-bill appears on signs in the historic section of St. Francisville, touting the neighborhood's status as a bird sanctuary. In this setting of graceful antebellum homes and plantations amidst live oaks draped with Spanish moss, one cannot help but feel nostalgic for the era in which ivory-bills flew through the swamps near St. Francisville.

After our lunch, Keith Ouchley and I headed to Cat Island National Wildlife Refuge, located just outside of St. Francisville, to talk about ivory-bills among some amazingly large cypress trees. Cat Island was purchased by the Nature Conservancy and transferred in October 2000 to the federal government. The number of large cypress trees still standing at Cat Island make this an impressive bottomland forest. Ouchley pointed out that these trees were not left standing because someone thought their size or beauty warranted preservation, but because they were hollow or crooked and so deemed useless by the former timber-industry owners. Many of these huge trees were standing when Audubon visited the area and ivory-bills still existed locally.

As I walked around Cat Island with Ouchley, I could sense the pride he takes in the area; it was his organization that was responsible for forever protecting this bottomland forest. In the time he has spent studying and working here, he has come to know well the area's plants, trees, and birds. While I scanned the ground for snakes, Ouchley moved through the forest with great ease, as if he were in his own backyard.

Ouchley, while intrigued by sightings such as the 1999 Pearl River report and others, said that he was skeptical that any ivory-bills survived. This was disappointing to me at the time. Given Ouchley's expertise in bottomland forest and avian ecology, I had hoped he would tell me that he believed there were ivory-bills somewhere in the swamps that he has studied. If an expert like Ouchley believed that ivory-bills survived, it would have lent persuasive weight to the case for the bird's existence. Although Ouchley didn't outright contest the existence of the ivory-bill, he did present a couple of reasons for being skeptical that remain as valid today as they were before the bird was rediscovered.

First, he disputed the notion that the birds are extremely shy, making it next to impossible for anyone to get a photograph. "There is no indication given by past writers or naturalists that these birds were exceedingly shy. Even Tanner wrote about the fact that ivory-bills actually become accustomed to humans pretty quickly," Ouchley said. While he admitted that hunting may have taught the birds to be more wary of humans, he found it curious that there have been so few photos taken of the bird during the past fifty years. Surely if the birds still existed, he said, they would have become less shy in some areas, allowing for photographs.

Next, Ouchley said he believed that from the standpoint of forest composition, the ivory-bill's habitat has been forever altered due to past massive logging. According to Ouchley, almost the entire bottomland forest habitat along the Mississippi River has been logged. "There is essentially no virgin forest left of any size," he said. Of the 24 million acres that once existed in the Mississippi River alluvial valley, only 2,000 acres are considered virgin timber today. That that may be true, I responded, but there are many areas, some protected, that contain large tracts of bottomland forest that could provide habitat for the ivory-bill. While he admitted that some large stands of mature bottomland forests do exist today, he countered with an observation based on his own research: in many areas, such as Cat Island, the tree species composition is radically different from what existed prior to logging.

The changes in tree species composition make many areas unsuitable habitat, according to Tanner's study in the Tensas River National Wildlife Refuge area. Tanner reported that the ivory-bills in the Tensas area overwhelmingly sought out their favorite diet of beetles and grubs in the dead branches of both the sweet gum and the Nuttall's oak. According to Tanner, bald cypress is far down the list of trees important in providing the ivory-bill food. And today, in areas such as Cat Island, the large, old, hollow bald cypress trees that were spared by loggers are what dominate the upper canopy, with the sweet gum trees present being quite small. Could the surviving ivory-bills have adapted quickly enough to the changed tree species composition, altering their foraging habits and becoming more dependent on other species as sweet gum trees disappeared?

While it may be possible, said Ouchley, it is not likely that modern bottomland forests dominated by cypress and tupelo gum would provide very good habitat for ivory-bills, even if these bottomlands contained mature trees. The large trees certainly look impressive—Cat Island, for example, is home to the largest bald cypress tree in the world—but if they are the wrong species, they do not necessarily make good habitat. This remains a point of interest today among foresters and ornithologists concerned with the ivory-bill's future.

On our walk through Cat Island that day, prior to the Arkansas report, Ouchley said that despite his skepticism, he would not totally discount the possibility that an ivory-bill or two still lived. He repeated that the habitat has changed, making it hard for the scientist in him to believe the bird survived. And even if it did survive, what would the habitat change mean to its future? But it also was hard for Ouchley to ignore all the rumors and claimed sightings. He specifically pointed to a couple of sightings in the 1980s that were made by experts in the Atchafalaya Basin, Tommy Michot and Bob Hamilton, whose interviews follow. And Ouchley told me about a rumor that the U.S. Fish and Wildlife Service was aware of the specific location of an ivory-bill or two in Florida but had kept the information secret. So while these tidbits of information kept Ouchley hopeful, the massive changes to the bird's habitat nagged at him.

After the Arkansas and Louisiana sightings, Ouchley and I met on several occasions in the summer and fall of 2005. Ouchley's perspective had changed, but he remained very cautious about the ivory-bill's prospects. He said he realizes that, like so many others, he had overestimated the importance of virgin timber to the ivory-bill and underestimated the bird's ability to adapt to changing habitat. Like others, Ouchley had based his original opinions on Tanner's study from the 1930s, the most detailed study of the ivory-bill that exists.

Today, Keith Ouchley heads up the State of Louisiana's Ivory-billed Woodpecker Task Force, which is charged with identifying search areas and carrying out ground reconnaissance in those areas. He admits he was surprised by the rediscovery in Arkansas because his hope that the bird survived had been slim. But now he says we owe it to the ivory-bill to conduct a more thorough search, to find out exactly what is out there. He cautions that we are not presently managing most of the second-growth bottomland forests for old growth. Instead, he says, we are managing it for game animals like deer and for timber board feet. If the ivory-bill is to recover, Ouchley cautions, forestry practices need to be reevaluated within many protected areas in the South. This is a theme that was reiterated by most of the professionals interviewed for this book. Ouchley stresses that even though the bird exists, there is no time for complacency. We need to alter our forest-management practices and develop a better understanding of what seems to be the ivory-bill's rapid adaptation to the changing bottomland forests.

Ouchley remains optimistic after the 2005–6 Louisiana search, but he is frustrated that it didn't provide hard evidence of the ivory-bill's presence. Having begun with so much promise, the search was derailed by Hurricanes Katrina and Rita, followed by the helicopter crash and fire near the search locations. During our last meeting, in June 2006, Ouchley said he believes the ivory-bill exists in Louisiana, but he again expressed concern about its recovery prospects. As we parted, Ouchley repeated that we owe it to the birds to do as much as we can to try to find and protect them. Given his history of working in and around ivory-bill country, I have no doubt Keith Ouchley will do everything in his power to support that cause.

Kelby Ouchley: U.S. Fish and Wildlife Service

Kelby Ouchley is Keith Ouchley's older brother. An official with the U.S. Fish and Wildlife Service and manager of the Black Bayou Lake National Wildlife Refuge in northern Louisiana, Kelby is also well versed in ivory-bill lore and history. He was, in fact, instrumental in bringing James Tanner back to the Tensas River National Wildlife Refuge in 1986 after the federal government opened the Tensas refuge. Surprisingly, Tanner had not stepped foot in the area since the early 1940s. Because of the destruction caused by logging in the 1940s, Tanner was reluctant to accept Kelby Ouchley's invitation, even some forty years after the area's forests had been felled. Because Tensas had been designated a national wildlife refuge, however, Tanner agreed to revisit his former field site. Tanner's visit gave Kelby Ouchley the privilege of interacting with the godfather of ivory-bill research in the very area that was thought by many to have housed the last ivory-bills on earth.

As I drove north to Black Bayou to meet with Kelby Ouchley in December 2003, I passed through many areas that were once home to ivory-bills. In fact, the path to Black Bayou from Baton Rouge—which takes one through St. Francisville and farther north near the Tensas River National Wildlife Refuge—is a trip through ivory-bill history. The trip also takes one through the Mississippi River Delta lowlands, which was prime ivory-bill territory until it was transformed for agriculture.

As interesting as it was to contemplate Tanner's and Audubon's adventures in search of ivory-bills, I also found the route quite depressing because so little forest remains along Louisiana State Highway 15 in the Louisiana Delta. Cotton gin after cotton gin tells the story of what became of the bottomland forests. Today, only barren fields remain in what was once one of the richest ecological landscapes in America. And the human landscape is similarly degraded and bleak, consisting mostly of hardscrabble towns that have seen more prosperous days. The days when cotton was king in this countryside are long gone, leaving behind empty storefronts and decaying mobile homes.

I frankly wasn't expecting much from Black Bayou Lake National Wildlife Refuge after driving through these depressed towns and cutover forests. But when I reached the refuge, just on the outskirts of Monroe, Louisiana, I was pleasantly surprised. The scenery around the refuge's main lake was beautiful. It was a cold day, with just a dusting of snow on the ground. As a suburban protected area that is much more accessible than most, Black Bayou no doubt plays an important role in educating the public about the importance of bottomland habitats.

In this interview prior to the 2005 Arkansas report, Kelby Ouchley—

like his younger brother, Keith—was skeptical of the reports of recent ivory-bill sightings. In fact, skeptical is not a strong enough adjective. At our first meeting, Kelby unequivocally said that not only was the ivory-bill extinct but that, in his view, it had not even survived the destruction of Tensas back in the 1940s. Of all the experts I interviewed for this book before the Arkansas report, Kelby Ouchley seemed to hold the strongest feelings that the ivory-bill no longer existed. The Arkansas report has softened his stance a bit, but not entirely.

Kelby Ouchley articulated very sound reasons for his opinion. Like other skeptics, he claimed that the ivory-bill was not a shy bird, so if it did exist—even just a couple of individuals—someone would have heard or seen them. He also said that the ivory-bill was an old-growth forest specialist and that virtually no bottomland old growth had survived the logger's saw. In support of Tanner's assessment that ivory-bills went extinct as the forests in the Tensas area were cut in the 1940s, Kelby Ouchley pointed to the fact that there simply was not enough bottomland forest habitat remaining after that point to support an ivory-bill population. Without the ample supply of dead and dying trees that the old-growth forests provided, Kelby Ouchley said, the ivory-bill could not have survived.

And even if a few birds had survived somewhere in the South and had managed to find each other and breed, Kelby Ouchley said, they would have been doomed from a genetics standpoint. He cited the red-cockaded woodpecker as a modern example. Because these diminutive woodpeckers are sometimes limited to isolated pine "islands" today, inbreeding has taken its toll. It took only a couple of generations before serious genetic problems such as crossed bills appeared in the bird. And, Kelby said, there are many, many more red-cockaded woodpeckers around than ivory-bills, making the red-cockaded woodpecker more genetically fit.

Regarding the reported pre-Arkansas sightings of the ivory-bill, Kelby Ouchley said he does not believe most individuals outright lie about seeing the bird; he believes they make honest mistakes. He said he thinks people want to see an ivory-bill so badly that they convince themselves that they actually did see the bird. By way of example, he points to the numerous unconfirmed reports of "black panthers" in Louisiana. He has talked with many individuals who swear they have seen black panthers, yet there has never been a single shred of verifiable evidence that they exist. He did receive one report from a couple of hunters who claimed to have captured a sighting on videotape while deer hunting. His initial review of the recording led an incredulous Ouchley to believe that the animal in the video was truly a panther. However, when the tape was analyzed frame by frame, it became clear the animal was not a panther but rather a coyote loping through a pasture.

When I interviewed Kelby Ouchley about the Arkansas reports on the telephone and via e-mail in the summer of 2005, he said he remains cautious about their accuracy. Like some others, he is not even sure the bird in the video is an ivory-billed woodpecker. Nevertheless, Kelby claims to feel more optimistic now about the chances of ivory-bills being alive. "The door isn't closed on the ivory-bill quite yet, but, given the large-scale habitat changes, I am not totally convinced that it has a future," Kelby said. But he also acknowledged that "we were obviously wrong about its habitat needs if the ivory-bill is alive in Arkansas." He said he wonders why it has taken so long to find the ivory-bill if it does exist because the White River area is used a great deal by hunters and fishermen. It seems to him that some of them would have seen and photographed the bird before, especially if Tanner was correct about the bird not being particularly shy.

Kelby said that the Arkansas reports alone do not yet warrant large-scale changes to forestry-management practices in wildlife refuges, but if more evidence of the bird is produced, then certain forest-management practices, such as promoting more old growth, will need to be revisited.

John Fitzpatrick: Cornell University

John Fitzpatrick's office at Cornell University in Ithaca, New York, is far from the historical range of the ivory-billed woodpecker. Yet the director of Cornell's internationally known ornithology laboratory is as drawn to the ivory-billed woodpecker as any southern birder or ornithologist I met while writing this book. Since the Arkansas film clip was released in 2005, he has been the figure from that search team most often in the public eye, and the one most often quoted in response to criticism of the team's sightings.

During a telephone interview in August 2003, Fitzpatrick said his fascination with the ivory-bill has been driven by many factors. First, the beauty, size, power, and the overall grand presence of the ivory-bill have drawn Fitzpatrick, like so many naturalists before him, to the bird. He mentioned that the ivory-bill was Audubon's favorite bird because it was such a magnificent and powerful creature. And Fitzpatrick said that his role as director of Cornell's ornithological laboratory makes him feel a historical connection with the study of the ivory-bill. James T. Tanner earned his Ph.D. from Cornell by studying the ivory-bill in the Singer Tract in the early 1940s. It was also Cornell's ornithology laboratory that recorded the ivory-bill drumming and calls in the Singer Tract in the 1940s. Even compared with the more recent recording, these ghostly recordings still provide the best audio evidence of an ivory-bill. You can visit the laboratory's Web page and listen to this original recording (www.birds.cornell.edu/ivory). Fitzpatrick said he feels a great sense of pride in Cornell's long involvement with ivory-bill research and searches, and rediscovering it has deepened the university's relationship with the bird. To Fitzpatrick, the ivory-bill is a tragic symbol of the human abuse of the southern bottomland forest ecosystem. You can sense deep frustration, if not anger, when he describes the rape of the bottomland forests as a "catastrophic failure" of U.S. conservation and resource management in the early twentieth century.

Before the Arkansas search and rediscovery, the scientist inside Fitzpatrick told him that the ivory-bill was "probably extinct." But he also held out some hope. The fact that many areas previously inhabited by the ivory-bill now supported mature bottomland forest and were protected gave him some hope the bird survived. He told himself that if a few birds had survived the first half of the twentieth century, when logging had eliminated most of the remaining old-growth forest in the South, maybe a few ivory-bills could have hung on until today. While the doubts nagged at him, he continued to look because, in his own words, getting out and looking for the ivory-bill "is just freaking fun. To walk through forests where we know ivory-bills existed, sometimes just a few decades ago, is really amazing."

Back in 2003, Fitzpatrick said in a telephone interview that many areas in the Southeast such as the Pearl River area and the Atchafalaya Basin in Louisiana, the White River region in Arkansas, and the Apalachicola National Forest region in Florida could support ivory-bills. The realization that good ivory-bill habitat still existed made him and others believe there was sufficient reason to keep looking for the bird. When asked if he believed the 1999 report from David Kulivan, the LSU forestry student, or the many other reports that have surfaced over the years, Fitzpatrick was noncommittal. He said that Kulivan had provided details that only experts would be aware of, such as the shape and direction of the bird's crest, yet after the failed attempts to record the bird, Fitzpatrick is uncertain. "Hunters get up very early, sometimes they fall asleep. Maybe that's what happened [a dream]," he said. Yet Fitzpatrick acknowledges that the location in the Pearl River Wildlife Management Area from which Kulivan's report came had experienced a tornado two years prior to the report, damaging many large trees. These damaged trees would have attracted the beetle larvae that are the ivory-bill's main food source.

Fitzpatrick also pointed out reasons that some of the reported sightings might not be accurate. Everyone who claims to have seen an ivory-bill, he noted, is unfamiliar with the bird. Thus he or she must look it up in a field guide, and by looking it up after the fact, individuals may well transfer features of the drawing to their memory of what they saw in the field. He pointed out that no birder with a pair of binoculars has ever reported seeing the bird—until the Arkansas and Louisiana reports. While this statement is not completely accurate—several brief glimpses have been reported by trained naturalists or resource managers, including Tommy Michot—Fitzpatrick's point is well taken. It would be relatively easy to mix up features seen in the field for a few seconds with those on a printed page, especially after looking at the picture a number of times and as time passes. Of course, this is the very claim—that the features of a pileated woodpecker are being confused with that of an ivory-bill—that critics of the Arkansas ivory-bill film are making today (figure 8). Like most experts, Fitzpatrick acknowledged in 2003 that there was evidence that supports the belief that the ivory-bill continues to exist, but he emphasized then that there was also a great deal of contradictory evidence. One year later, Fitzpatrick's sliver of optimism paid dividends when he participated in the Arkansas search that led to the eventual filming of the bird and the announcement in *Science* magazine.

Even though Fitzpatrick had some doubts about the bird's existence, he developed a game plan in case one was sighted. And, based on what happened after the Arkansas search, he seems to have followed the plan closely. He said that, if at all possible, he would keep the sighting a secret until he was able to alert the proper state and local authorities so that they could implement a plan to limit the number of birders that he suspected would enter the area, seriously disturbing what might be the last ivory-bill on the planet. And he expected intense interest from more than just serious birders. Fitzpatrick told me that in his discussion of the ivory-bill issue with preeminent ecologist E. O. Wilson, both had agreed that a sighting would be big enough news to warrant a cover photo on *Time* magazine. While the bird did not make the cover of *Time*, the news of its sighting certainly received media saturation in 2005. Fortunately, even with the thorough media coverage, the White River National Wildlife Refuge was not overrun by birders.

While Fitzpatrick was uncertain about the existence of the ivory-bill at the time of the 2002 search, he does not see that search as a wasted effort. Like Van Remsen at LSU, he felt that it helped to make the later Arkansas search successful. The 2002 search offered his lab the chance to try out new technologies such as the recording devices used in the White River National Wildlife Sanctuary. Fitzpatrick said that he believes that the previous destruction of the bottomland forest and its most impressive avifauna resident also have provided an important lesson by giving researchers a better appreciation of the bird's bottomland forest home. "We should manage these impressive secondary forests in the Southeast as if the ivory-bill still existed," said Fitzpatrick in 2003.

In the time since the announcement of the ivory-bill's rediscovery, Fitzpatrick has been conducting a follow-up search in Arkansas and contributing his experiences and Cornell's technology to the Louisiana search. He is convinced that if the ivory-bill can survive in Arkansas and Louisiana, it might survive elsewhere. The Cornell-led team and Fitzpatrick specifically have also spent a great deal of time since the 2005 announcement defending their evidence. According to various search team members, there are no doubts among the group's members, including Fitzpatrick, that they saw one ivory-bill and heard more than one and that the birds still exist. They may not be in Arkansas—possibly having left using the Mississippi River as a corridor—but they do survive.

Louisiana State University

An academic center for past and present ivory-bill authorities, Louisiana State University, too, can claim a historical connection with the bird.

While writing this book, I was struck by how many individuals associated with LSU over the years have had some personal experience with the ivory-billed woodpecker. Professor George Lowery was the early ivory-bill lightning rod at LSU, and there continues to be a great deal of interest in the bird among current and retired faculty members and former students. Given that LSU sits on the eastern edge of the Atchafalaya Basin, this focus certainly makes geographical sense. In the nineteenth century, the grand bird is likely to have flown near the present location of the campus.

Vernon Wright

One recently retired LSU faculty member with a great deal of interest in the ivory-bill is Professor Vernon Wright. Of all the experts I interviewed, Wright was the most insistent that the ivory-bill survives, and his optimism dates to well before the news broke about the Arkansas search results. Vernon Wright is a respected scholar who has researched and published articles about statistical aspects of wildlife management, populations, and conservation biology. He is also an avid outdoorsman who paddles the bayous, rivers, and swamps in Louisiana. Although he is not an ornithologist, Wright is very familiar with the birds and swamps of Louisiana and southern Mississippi. I first met with Wright in his office at LSU in early July 2003, and we have had several phone conversations and e-mail exchanges during the past three years.

While paddling with Wright in the northern edge of the Atchafalaya on Two O'clock Bayou, I was impressed by how many birds he immediately knew by sight and sound. Anyone who paddles the Louisiana waterways cannot help but have the ivory-bill on his or her mind, especially if the paddler is a wildlife professor. But Wright was pulled directly into the ivory-bill controversy after his student, David Kulivan, reported seeing a pair of ivory-bills in the Pearl River Wildlife Management Area in 1999. In fact, Wright was the first person Kulivan told of his sighting. Wright is convinced that Kulivan did indeed see a pair of ivory-bills, partly because he himself believes that he has gotten glimpses of ivory-bills on two and possibly three previous occasions while paddling his canoe. While Wright never saw the birds as clearly as Kulivan, he claims to have seen large, crested woodpeckers flying rapidly, in a straight line—as opposed to the pileated, which undulates as it flies. Wright is not 100 percent sure he saw ivory-bills, but he asks, "What else could they have been?" His sightings occurred in the mid-1990s in Two O'clock Bayou in the northern section of the Atchafalaya and on the grounds of the Stennis Space Center in southwest Mississippi. Wright points out that both areas are connected to much larger wilderness zones: Two O'clock Bayou is near the Atchafalaya Basin, and the Pearl River Wildlife Management Area and Bogue Chitto National Wildlife Refuge are adjacent to Stennis, which is also quite large and has the potential to support a few ivory-bills.

After paddling Two O'clock Bayou with Wright, and later on my own, it seemed feasible to me that ivory-bills could be in the area. The area is largely leased by hunting clubs that have kept the mature bottomland forest intact. Because the land is leased, no one other than club members enters it. So no one has searched for the ivory-bill in that area. While it is mainly second growth, the trees now are quite mature, with some resembling old growth. One huge sweet gum caught my eye because the dead, protruding branches at its top resembled some of the trees Tanner photographed in the 1930s that were used by ivory-bills in the Tensas area. Wright and I floated quietly in the shade, watching the sweet gum and both wishing an ivory-bill would appear again, recreating Tanner's photo. Instead, we were treated to an aerial show by a group of swallow-tailed kites that were catching dragonflies midair in the oppressive afternoon heat. Paddling the bayou that day once again reminded me just how hard it would be to get a good photograph of an ivory-bill—or even of a perched bird—inside the green wall of vegetation that lines the banks.

Unlike more skeptical experts who concede that a few old remnant individual ivory-bills may still be alive simply because the birds are long-lived (sometimes living more than twenty years), Wright said he believes that there are a few tiny populations of ivory-bills, and that these birds are breeding. Therefore, the reports of birds from around Louisiana, according to Wright, might be describing sightings of young ivory-bills that are dispersing, looking for new territory. Given the sightings in Arkansas and Louisiana, it seemed to me that Wright's theory could be accurate. Surely these birds have bred and dispersed over the last few decades. How else would a bird have been found in Arkansas, the same place that was cut over earlier in the twentieth century?

When I spoke with Wright after the rediscovery—on the telephone in May 2005 and then in June 2006 at CC's Coffee House in Baton Rouge—he felt to some degree vindicated in his former optimism about the bird's survival, of which some other experts had been critical. More important, though, he was happy that the bird survives. He said that he truly believes that the skepticism of many ornithologists about past reports has often led to missed opportunities to find the bird. Wright is an outdoorsman who feels a bond with the hunters and fishermen who have reported seeing ivory-bills. He said that many in the birding and scientific communities have been too quick to dismiss their sightings, partly due to simple snobbery. After witnessing the barrage of criticism fired at his former student David Kulivan after his 1999 sighting, Wright came to believe that weekend birders who look down on hunters and fishermen are shortsighted, even foolish. Wright said that he strongly suspects that there may be more ivory-bills out there than previously thought. Although he acknowledges that the species is critically endangered, he also believes that there are enough birds to breed and disperse, especially in Louisiana.

Bob Hamilton

Wildlife professor Bob Hamilton has also spent many years at LSU and claims to have both seen and heard the ivory-bill. Hamilton, now retired from LSU, has been described by some of his former students as being a very serious, ethical, and expert ornithologist, one too principled to exaggerate what he has seen. For example, former student Keith Ouchley, whose interview appears earlier in this chapter, described Professor Hamilton as "probably the best birder I know. If he says he saw a certain bird, I believe him." While this may not seem like such a rare characteristic for a former wildlife professor, I make this point because Hamilton claims to have heard and seen not only an ivory-billed woodpecker, but also a Bachman's warbler (*Vermivora bachmanii*).

The story of his Bachman's warbler sighting provides insight into the ethical and scientific side of Hamilton. According to Hamilton, he was counting birds along a transect in the Tensas River National Wildlife Refuge in the early 1990s at specific time and space intervals. At a certain point, he thought he identified a Bachman's warbler. However, even though he knew that the bird was thought to be extinct, he was reluctant to leave his transect location because its data depended on his consistency. After some internal debate, he left the transect and followed the warbler, but the dilemma this posed for him illustrates his dedication to proper methodology. He observed the Bachman's warbler for about ten minutes. If his identifications are accurate, Hamilton may be the only person alive to have seen both an ivory-bill and a Bachman's warbler. Given Hamilton's reputation and conduct, I and many others consider his claims to be legitimate. He has never been vocal about his reports, instead remaining largely in the background during the recent search for and discussion of the ivory-bill. Hamilton participated in the 2005–6 search in Louisiana, but he remained modest and quiet about his role.

I had the opportunity to meet Hamilton and discuss the ivory-bill twice, once in August 2003 in Baton Rouge, where he lives, and the second time when we drove to Cat Island National Wildlife Refuge in October 2003. Going into the field with experts always interests me more than a conversation in an office or coffee shop because I gain a greater sense of their experiences with nature in general and the ivory-bill specifically. While Hamilton claimed not to be an obsessive ornithologist, he was keenly aware of and interested in any bird calls we heard while walking through the cypress forest at Cat Island.

Like some others who have reported seeing an ivory-bill, Hamilton claims to have seen one while driving across the Atchafalaya Basin on Interstate 10 near Rama, Louisiana, just west of Baton Rouge. While he could not remember the exact year, he claims he saw the bird sometime in the early 1970s, probably 1973 or 1974. This sighting, like so many others, was fleeting. And while most sightings of an extremely rare bird while traveling sixty miles per hour might be dismissed, Hamilton's reputation lends a great deal of credibility to the story.

Several years later, in about 1978, he heard the ivory-bill's distinctive double-rap and a loud, nuthatchlike call while in the Atchafalaya Basin, both sounds seeming to have been made by the same bird, according to Hamilton. He heard these sounds in the same area from which a couple of other solid reports of ivory-bills came during the late 1970s, 1980s, and even 1990s. According to a 2003 e-mail from Hamilton: "As I was walking down the trail on my way back to the boat, I heard a call to the north of the trail. This was the nuthatchlike call that we know and I had reviewed before my trips. I thought it was from an ivory-bill, but I was not 100 percent sure. It was not exactly like the one on the tape. I am very good with bird calls, and I am positive I had never heard that call before. The bird stopped calling, but I soon heard the tap-tap at the cypress tree on the south side of trail. The distance I estimate between the presumed location of call and the cypress-tree tap-tap was at a maximum fifty meters. Dr. Newman [a former professor at LSU] was with me and was quite excited about our adventures. He did not know bird calls like I do, but he was pretty convinced [that it was an ivory-bill]."

This incident as well as some other reports of ivory-bills nearby prompted him to venture out on his own into the Atchafalaya Basin and search the area from which the double-rapping had originated. However, after ten days in the swamp and unpleasant encounters with an abundance of stinging nettles, he decided to end his search. Hamilton's experiences in the swamp illustrate the challenges posed by the terrain in ivory-bill habitat. If a dedicated ornithologist like Hamilton was forced to leave, how many weekend birders would last even a day fighting the bugs, snakes, and stinging nettles? While most of the birds that birders seek can be found on the fringes or edges of bottomland forests, that is not true of the ivory-bill. And it seems unscientific, at best, to deny the presence of a species unless significant time has been spent searching its habitat.

Hamilton has followed the debate about the ivory-bill with interest. While he believes he saw and heard the bird decades before the 2005 Arkansas report, he did not until quite recently believe that it still existed in Louisiana. His participation in the 2005 Louisiana search, in which he may have heard the distant calling of an ivory-bill, has made him more optimistic about the presence of the bird in the Bayou State.

Prior to that experience, Hamilton felt, like many others, that the birds he had seen in the 1970s and even the early 1980s were remnant individuals that had hung on in remote corners of bottomland forests and swamps such as the Atchafalaya. And today he still considers that a possibility. He said he remains unsure about other past sightings. For example, while he does not believe that David Kulivan's account of a 1999 sighting in Pearl River was fabricated, he does not believe a breeding pair of ivory-bills exists in the Pearl River area today. As we walked around Cat Island, under towering cypress, Hamilton talked about the sheer odds of a bird, any bird, surviving until today given the geographic isolation imposed by habitat fragmentation and general habitat destruction. His sound reasoning echoed that of Kelby Ouchley.

Hamilton also initially disputed the notion that the ivory-bill, if it still exists, would be hard to find, although his participation in the Louisiana search softened that opinion. Prior to the Arkansas report, he said he believed past reports such as Tanner's that the ivory-bill is not a reclusive or secretive bird, and that the large and loud woodpecker would easily be spotted. Because Hamilton spends a great deal of time in the Atchafalaya Basin fishing and birding, to him it is not a distant, impenetrable landscape. The idea that the bird could remain hidden, evading capture by the cameras of the thousands of hunters and fishermen who venture into the basin, struck him as preposterous.

Although Hamilton was happy and excited after the Arkansas sightings, he remained cautious. Asked in May 2005 if he now could add the ivory-bill to his birding life list, he stated: "The news itself did not change my view much, and it did not change my life list at all. It does establish that at least one bird still exists, but it was not found where I probably encountered a bird before." Like others, Hamilton said he changed his views about the bird's shyness because the Arkansas bird was extremely shy and quiet.

"The extreme wariness makes me feel better at not being able to see the bird I was apparently so close to," Hamilton said. "This extreme wariness was somewhat surprising because the historic record indicates that ivory-bills were often quite noisy and conspicuous. On the other hand, Tanner found them sometimes secretive. Certainly if there were still any alive, they could not be very conspicuous because there is little (if any) habitat that is not visited regularly by humans. Of course there may have been, and probably were, some legitimate contacts by outdoorsmen that were not believed or investigated."

I also asked Hamilton if he felt somewhat vindicated now that he knows the chances are good that he really did see and hear an ivory-bill. He replied, "I feel somewhat vindicated because many of my opinions were confirmed. Primarily, I did not believe that a one- or two-hour excursion by bird-watchers was a good way to find the bird." Birders who do not want to get hot and wet or come into contact with snakes are unlikely to accomplish the goal of finding more birds, Dr. Hamilton believes. He said he realized long ago that in his searches for the bird, he needed to spend a lot of time actually in the swamp, not on the periphery, so he camps. "Camping takes a lot of faith if all you are doing is looking for a bird that might be extinct, or if not extinct might not be in the area you are searching at the time of the search. Modern remote sensors obviously make things a lot easier. I now chastise myself some for letting the inconvenience of nettles shorten my one serious attempt to find the bird. In my defense, I did not have funding, and my efforts were largely self-supporting," he said.

Although the rediscovery of the ivory-bill surprised Hamilton, he said he was not shocked that of all the potential locations, the bird was found in the Cache River National Wildlife Refuge. He stated that "after looking at the habitat situation, the White River region [including the Cache] may be the most logical place to find ivory-bills because the habitat seemed better there to me than other places I knew of. I am not familiar with the area where the bird was found, but I did take a trip to White River and talked to some staff there. Unfortunately, they discussed their forest-management program, [which] I thought was not appropriate for creating or maintaining ivory-bill habitat. I imagine the bird was found in an area that was not heavily managed."

Hamilton's comments shed light on a developing conflict: If federal and state authorities are going to start managing refuges to create and/or enhance ivory-bill habitat, major changes to forestry practices will have to take place—the call for changes in how we manage our bottomland forests was a recurring theme in my interviews with the experts. Hamilton asserted his belief that "the failure to manage for more climax or climax-like habitats is an enormous mistake, and the finding of the bird in a climax or unmanaged area may provide ammunition to change our [ecological] priorities."

As more reports come in, and as Hamilton gets more involved in the search for the bird in Louisiana, he has become more open to the possibility of a recovery for the species. But for Hamilton, as for many others, his optimism is clouded by the fact that habitat has been destroyed in the past and that second-growth forests continue to be poorly managed by state and federal authorities. In his view, the emphasis on producing more timber, not species diversity, in most bottomland-management areas continues to threaten any future for the ivory-bill.

James "Van" Remsen

Today, Professor James "Van" Remsen, the curator of birds at the LSU Museum of Natural Science, is the person whom those in the large Louisiana birding community most associate with the ivory-billed woodpecker. He was involved in both the 2002 Pearl River search and the 2004 Arkansas expedition. While Remsen has received a lot of publicity during the media coverage of the Pearl River and Arkansas searches, he has not sought to inflate his role in the search or to downplay the fact that the ivory-bill is a secondary area of investigation and research for him. He pointed out in our follow-up interview in June 2005 (as he had prior to the rediscovery announcement) that he and his lab are primarily focused on tropical ornithological research, not ivory-billed woodpeckers. Nonetheless, he said he was thrilled with the Arkansas rediscovery and with being able to play a useful role in recent ivory-bill searches. He has fielded calls about the ivory-bill for a couple of decades, so even though his own research has been focused elsewhere, his thoughts were never far from the ivory-bill.

Our first conversation, prior to the Arkansas sightings, took place in Remsen's office, which is in a back room of the museum, hidden behind cabinets containing thousands of bird specimens. Remsen described the three factors that he thought accounted for people's fascination with the ivory-bill: "First, it's a large and spectacular bird. It's the largest woodpecker in the South, and so naturally it would draw attention. The draw to the bird because of its size and beauty can be traced to indigenous peoples in the South and beyond," he said. The second factor, according to Remsen, is "the fact that people have seen the bird not that long ago. So it's not like an animal that vanished hundreds or thousands of years ago. Live people remember seeing the bird, so it makes it more real for others." And last, and perhaps most important, said Remsen, "the ivory-bill is a symbol of our lost natural heritage in the South." This observation that the ivory-bill's disappearance was symbolic of the destruction of the wild South was a comment I heard frequently from those interviewed for this book.

Before the Arkansas rediscovery, Remsen was cautiously and quietly optimistic about the continued survival of ivory-bills. He was not a loud cheerleader; that is not his style. But he had received enough good reports over the years to make him hopeful that a few birds probably survived. Since his personal research did not involve woodpeckers, he did not often follow up on the reports that regularly came into his office at LSU, with the exception of the Pearl River search, which he helped coordinate and lead. While he felt fairly certain that a few past reports were legitimate—such as Fielding Lewis's photos and various stories from the Franklin area—until the Pearl River expedition he did not have the time or resources to organize any formal searches. He had ventured into the Atchafalaya Basin on a couple of occasions to track down a particularly good story, but he now admits that weekend searches won't find the ivory-bill. Even at the time of his short trips into the basin, he wasn't overly optimistic that short searches would produce any results. And given the slim chance of finding a bird, he has never encouraged any of his graduate students to pursue the topic, even after the Arkansas sighting. That, however, has more to do with the realities of academic life, where field support and research time are in short supply, than with Remsen's attitude toward the rediscovery.

Today Remsen is quite optimistic that other birds exist, particularly in Louisiana. During interviews in June 2005 and June 2006 and in several e-mail exchanges during the past two years, he reiterated his confidence not only in the controversial Arkansas video but also in the likely survival of ivory-bills In Louisiana. While he felt comfortable stating, after the 2002 Pearl River search, that the area contained no ivory-bills, he now freely admits mistakes were made and that the Pearl River Wildlife Management Area should be completely reexamined. For example, Remsen believes that the audio-recording devices did not cover nearly enough territory, and that the search team was not secretive enough. "While the Pearl team wore camouflage, and we were quiet, based on how shy the bird was in Arkansas, we totally underestimated just how quiet we needed to be. We were basing that opinion on past reports that claimed the bird was noisy, easily detected, and not especially shy," he said.

Remsen pointed out problems with past studies. He believes we may have put too much faith in Tanner's work in the 1930s, given that the study involved a single, small, and stressed population of ivory-bills whose forest was disappearing around them. Remsen was not critical of Tanner's work per se; instead, he pointed out that it might be a mistake to assume that all ivory-bills act the same way as those Tanner studied in the 1940s. The ivory-bill "crowd" is small, Remsen said, and because

Tanner authored the key study and was apparently a very nice man, few people come right out and state that he was wrong. According to Remsen, behavioral changes may have occurred in the ivory-bill since Tanner's study due to habitat differences and human impacts. Just as turkeys and ducks have developed behavioral changes in response to hunting pressures, ivory-bills may have become more secretive to avoid hunters. Remsen also said, in our first conversation in the summer of 2003, that "we need to rethink the habitat requirements," meaning that it might be wrong to assume that ivory-bills can survive only in huge expanses of old-growth timber. Remsen stated: "Woodpeckers in general are very flexible [in terms of habitat]. Nest sites [trees] and food are the key." He said that if a forest, even a relatively young forest, has enough dead and dying timber to provide nest trees and food—especially the grubs found under the bark of dead and dying trees—then the area could potentially house ivory-bills. Remsen is in no way stating that ivory-bills might be found in young, cut-over forests, but rather that it might be worth taking another look in areas where some old growth remains, as well as in large expanses of mature second-growth forest where there has been a history of sightings.

Remsen readily admits that what he found in Arkansas convinced him that he and many others had been wrong about certain aspects of the ivory-bill. He was correct, however, when I asked him in 2003 about sites that might house the birds. He stated then that the White River area was the first place he would look, given the quality of the forest. He also mentioned the Apalachicola National Forest and the Atchafalaya Basin.

For Remsen, the most exciting and important aspect of the Arkansas news and the reports of birds in Louisiana is that we have an opportunity to write another chapter for the conservation of both the ivory-bill and the bottomland forests. The book is not closed on the ivory-bill, and the recent sightings may help alter forest-management practices and how society views bottomland forest habitat.

Chuck Hunter: U.S. Fish and Wildlife Service

Chuck Hunter has a long and somewhat convoluted title. He is chief of the Division of Planning and Resource Management for the Refuge Program for the U.S. Fish and Wildlife Service in the Southeast Region.

While the term "ivory-billed woodpecker" does not appear in his title, Hunter has been spending much of his time since the Arkansas rediscovery on issues pertaining to the ivory-bill such as recovery, consultation, and outreach. He is essentially on the front line of coordinating the federal end of ivory-bill management, protection, and recovery.

Hunter's story is one of an early fascination with the ivory-bill remarkably similar to my own. As a young boy, Hunter had a keen interest in birds and rare animals. At an early age, he already owned birding field guides and spent many hours outside identifying local birds. In 1969, near Jacksonville, Florida, Hunter saw an unfamiliar large woodpecker. At the time, he thought perhaps he had seen an ivory-bill. This event led Hunter to contact the local chapter of the Audubon Society in Florida's Duval County. After that, he was hooked on birds. Throughout his school years, he participated in Audubon's Junior Program, which brought kids out on camping and birding trips. Eventually Hunter led these trips himself, introducing other kids to the wonders of nature that he had experienced years earlier. Hunter's lifelong interest in the ivory-bill, triggered by an event in 1969, has come full circle in his coordinating the federal government's efforts to recover the ivory-bill.

Even before the Arkansas rediscovery, Hunter was optimistic that somewhere in the Deep South a few ivory-bills survived. Perhaps his interest was driven by his childhood fascination with the ivory-bill; perhaps it was based on the fact that over the years he has received solid reports of ivory-bills from various locations in the South. Either way, Hunter never lost hope that the bird would again be found.

In a telephone conversation we had in early 2003, Hunter mentioned the White River area in Arkansas as a potential location for ivory-bills. He said he had received some solid leads from Arkansas, Florida, and, of course, Louisiana. Like many others, he expressed concern about the long-term prospects for a very small ivory-bill population, but he still remained optimistic about finding some birds. The fact that the ivory-bill had disappeared and then reappeared decades later gave Hunter hope that it would appear again. During our conversation in the summer of 2005, following the Arkansas sightings, Hunter could hardly contain his excitement. For him, the fact that the ivory-bill had been found and the fact that he was playing a role in its possible recovery was the realization of a lifelong dream.

In the following e-mail message to me from the fall of 2005, Hunter described the behind-the-scenes timeline regarding the Arkansas rediscovery. What a thrilling and nervous time that must have been for someone who has followed the trail of the elusive ivory-bill for so long.

> We were notified last spring (2004) that there had been some interesting sightings and it was inquired whether a more thorough and organized search would be allowed on the refuge (because it was a federal location), but kept out of the public eye as much as possible. We arranged for special use permits and ground rules for work on the refuges and similarly the state of Arkansas did the same on their wildlife management areas involved. While the search team did their thing, a number of us worked behind the scenes to make sure the search went smoothly and started working on contingency plans in case confirmation resulted.
>
> We (the U.S. Fish and Wildlife Service) did not know the extent of their evidence until early April of this year. We had instructed the search team that during the search proper they should not come to us until they felt they had publishable evidence. To the public, any questions about what these outsiders were doing was responded to by saying they are doing a bird inventory. When late March was coming on I called Scott Simon, State Director The Nature Conservancy in Arkansas and who was the nerve center for the search and the behind the scenes team, to suggest that with nothing coming to us so far (i.e., nothing that constituted confirmation), we should just get together sometime in May or June and review where we were with the search, etc. He strongly suggested we should meet sooner, like early April, so that we could be out in front of any rumors that might start after the search team dispersed by late April. We settled on the date April 7, when the video emerged and other evidence, plus a draft paper for *Science*. This was a shocker, but the search team had technically not held out on us; they had followed my instructions to the letter.
>
> They had a video as we now know from April of last year, but no one on the search team was ready to stake their careers on just the video and the sightings that had been tallied. By the following year, they had more credible sightings, audio evidence from several

widely dispersed areas, and of course this video. Still that was not considered enough to cross the threshold we had established for announcing their findings to the Service. As spring sprung and leaves started to emerge, they had not yet conducted the detailed analysis of the video. It was becoming clear that this might be the key piece of evidence that they were likely going to get so best to get in the field and . . . conduct measurements, and figure out all the potential critical angles that would eliminate [the] pileated woodpecker as the bird in the video. As you know, the process of the scientific method is to disprove alternatives, not to prove your hypothesis. So they set out to attempt disproving that the bird in the video could possibly be a pileated, weird plumage or otherwise. The weight of the evidence did in fact disprove the possibility that this was a overgrown, overly white pileated Woodpecker. So with the only alternative explanation eliminated, voila, what else could this be?

The search team had already submitted their article to *Science*, which was pending approval based on editing boards review. We were planning for a mid-May announcement coinciding with a release of the paper on *Science* Express [Web site for *Science* magazine] (again pending final approval) and an NPR *Radio Expeditions* piece on *Morning Edition*. Leaks became uncontrollable by Tuesday April 26 and the decision was made to announce on April 28. Amazingly, the NPR piece came on at around 7 am Thursday morning, the *Science* Express paper became available to Internet users later that morning as Secretary Norton and all the other dignitaries were making their formal announcement in Washington. Just like it had been planned that way for months (instead of two days). And where was I the day before?—at Tensas National Wildlife Refuge, the former Singer Tract, where the last confirmed Ivory-bills occurred in 1944. Sixty-one years later I was part of a biological review team helping to chart the future management of this refuge, with a pair of birds staring at me from the visitor center display case, with a bunch of people (all trusted colleagues) that I couldn't talk to about any of this. I'm sure they might have imagined something was up with all the phone calls I was taking from the refuge manager's office behind closed doors, but all in good time the events from the past year were revealed and then folks started to think about the future.

The rediscovery has forced Hunter as well as many others to rethink many issues concerning the ivory-bill. For example, he mentioned that habitat requirements of the ivory-bill may not be as dependent on old-growth forests as previously thought. The bird in Arkansas was using some forests that were quite young—it did not confine itself to only the old timber in the Cache. Because most of what is discussed and cited about ivory-bill ecology and behavior is based on only one study (Tanner's work in Louisiana in the 1930s), we may not know as much about this bird as we thought. In other words, Hunter said, we may have placed too much stock on a single study of a group of birds whose behavior was likely altered by the environmental stresses and forest degradation going on around them. And, like many other experts, Hunter said Tanner was far too dismissive of areas that may have still housed ivory-bills in the 1940s and later. As he pointed out, although most of the White River area in Arkansas was heavily logged during the twentieth century, the ivory-bill somehow hung on. The White River region certainly contains some areas dominated by ancient trees, but much of the area houses second-growth forest.

While Hunter is extremely excited about the rediscovery and the prospects of finding more birds, he knows many challenges also lie ahead if a recovery is to take place and if more birds are to be located and protected. According to Hunter, the most critical and large-scale challenge right now is whether we have the time and resources to find more birds. As the Arkansas rediscovery has proven, weekend searches are unlikely to turn up much, he said. The Arkansas team spent tens of thousands of hours in the field. Hunter believes that that kind of long-term, systematic search is what is needed in areas that are deemed most likely to house the ivory-bill. Other areas that need intensive searches are the Atchafalaya Basin and Pearl River area in Louisiana; the Big Thicket region in east Texas; the Santee Swamp in South Carolina; and the Apalachicola National Forest region in Florida, said Hunter. But there have just been too many reported ivory-bill sightings in those areas for him to believe they don't exist. Where there's smoke, he said, there's probably fire.

Another challenge Hunter described is to better understand the ecological dynamics of ivory-bills and how best to manage their habitat in a recovery project. Should forests simply be left alone? Should trees be singled out and girdled to produce food and nesting locations? Should other fauna such as beavers that flood forests, leading to some tree mortality, be allowed to carry on? This last point exemplifies some of the challenges that lie ahead in crafting a recovery plan. Beaver populations are often controlled on state and federal lands because they flood large areas of forest, eventually altering the species composition of the flooded forest (King, Keeland, and Moore 1998). Since dead and dying trees are critical for ivory-bills, it might seem that beavers would be beneficial. But this isn't a simple either-or situation (either leave beavers alone or harvest them). Too many beavers ultimately could result in poorer ivory-bill habitat because of beavers' propensity to completely alter the hydrological regime and create buttonbush swamps. On the other hand, limited beaver damage could improve some habitat, especially in the short term. Management polices will have to be fluid.

These are just some of the questions that will have to be addressed as a recovery plan unfolds. And as if time, money, and ecological questions aren't challenges enough, there are potential problems involved in finding a bird on private land. Hunter said that he realizes that if a bird is found on private land—and many of the most reliable recent reports of ivory-bills have come from private lands, as in the case of the recent Louisiana sightings and formal search—the situation could be tricky. Landowners, especially those who cut timber on their property, may fear losing access to certain resources, which often translates into losing money. Hunter understands that these fears may seem legitimate initially. However, he also stated unequivocally that private lands that house ivory-bills will not be seized. He said, "If landowners have woodpeckers on their land, they must be doing something right, so we are going to tell them, keep up the good work." Hunter went on to say that the federal government will create incentives for landowners to manage their lands in ways that would be more appropriate for ivory-bills and other species that depend on old forest. He does not foresee a confrontational relationship developing with timber owners. He acknowledges that some landowners are totally anti–conservation management. These landowners believe that they have the right to do whatever they want with their property. While this attitude could not be accepted if an area is known to contain ivory-bills, Hunter is genuinely interested in working with private landowners in a way that suits everyone's needs. Certainly state and federal authorities will play some role in future land-use decisions, but landowners should not fear that their land will be seized.

Today, in the post-Arkansas phase of ivory-bill search and rescue, Hunter is very optimistic about finding other small populations of ivory-bills. When asked to rank his optimism on a scale of one to ten, he would only say that he is very optimistic that there are other birds out there. Hunter also said that we may need to rethink our idea of proving the existence of ivory-bills "beyond a shadow of a doubt" before we accept their presence. Perhaps we set the bar a bit too high in the past, which may have led to accurate reports of birds being overlooked and ignored. Given that we now know the ivory-bill still survives, at least in Arkansas, Hunter believes that we cannot afford to risk losing other populations by ignoring the reports of local people, especially in areas where there is a history of reports.

Team Elvis South

The most enjoyable aspect of writing this book has been the people I have met and the places I have visited along the way. And no group of individuals associated with this project has been more enjoyable than the members of Team Elvis South—a humorous reference to the Arkansas search team, Team Elvis. This group is made up of Tommy Michot, Wylie Barrow, Dwight LeBlanc, and Garrie Landry. All are locals from the Lafayette area, and all are involved in one capacity or another with environmental and conservation research, education, and policy in Louisiana. Because they are local, all know the Acadiana region, including the Atchafalaya Basin, quite well. They all have extensive birding experience and advanced degrees in some aspect of ecology and/or wildlife-management issues, and all have been instrumental in providing leads on the ivory-bill.

I met this group through an early request for information about the ivory-bill on the birding list serve "LA bird," a forum for discussing ideas and theories about the ivory-bill's whereabouts. In the time since our initial contact, I have met with them on numerous occasions in Lafayette or in Franklin, Louisiana, along with Fielding Lewis, who sent the controversial ivory-bill photos to George Lowery in 1971. Inspired by Fielding's stories of the woodpecker, we often examined maps of the area pinpointing specific locations that held particular promise for ivory-bills. One such area was the Bayou Sale National Wildlife Refuge and surrounding areas, including the Patterson region. The following information is gleaned from our many meetings between 2003 and 2006, as well as from consistent e-mail communication among the group. For me, it is always so enjoyable and intellectually stimulating to share ideas and perspectives regarding the ivory-bill with other like-minded folks. And given that all the members of Team Elvis South grew up in the Acadiana region, they have great insight into the area's environmental issues, from the ivory-billed woodpecker to the constant rumors of big cats in southern Louisiana.

Tommy Michot

Tommy Michot, a wildlife biologist with the U.S. Geological Survey (USGS) and an accomplished musician and vocalist in his family band, Les Frères Michot, in Lafayette, may have the most direct connection with ivory-bills of anyone in the group because he believes he might have seen an ivory-bill near Duck Lake in 1981 in the southern Atchafalaya Basin. The Duck Lake area may be the first location targeted for extensive field searching by the Louisiana Ivory-billed Woodpecker Task Force. Michot's research, like that of the others, is focused on nonwoodpecker avian and coastal ecological issues, but his Ph.D. from LSU in zoology certainly qualifies him to speculate on the presence of ivory-bills.

Michot noted in an e-mail that his personal interest in the ivory-bill was jump-started by his grandparents' reported sighting of the bird: "I guess I first got interested in the ivory-bill around 1971 or '72. It just so happened that I was taking an ornithology class from Dr. Marshall B. Eyester around that time at the University of Southwestern Louisiana. I had an inquiry from my grandparents in Patterson about a big black-and-white woodpecker that they had seen on a tree near their wharf on the Lower Atchafalaya River. It fit the description of an ivory-bill, but they had no details that would have ruled out a pileated. I figured it was probably a pileated, but I mentioned it to Doc just for the hell of it. He replied that it just so happened that there was a recent sighting near Patterson, and so it was quite possible that it was an ivory-bill. I figured out years later that he was referring to the Fielding Lewis sighting and Lowery's involvement. Later, of course, my interest increased by an order of magnitude when Hankla and I had our sighting in 1981. Since then I have kept a file on it and tried to read whatever I could on the subject."

Michot's own sighting took place in 1981 while he and a colleague, David Hankla, were conducting a site inspection for a proposed oil-field canal for the USGS in the Atchafalaya Basin. As they turned the boat into a canal near Duck Lake, the woodpecker flew in front of their boat. They saw immediately that it did not look like the typical pileated woodpecker, with which both were quite familiar. White trailing feathers on its wing were clearly evident in flight. It landed on a willow tree some forty yards ahead of their boat. Hankla viewed the bird through binoculars while Michot quickly changed lenses on his camera in an effort to take a photo. As the bird flew off into the forest, the two landed their boat and set out on foot after the bird. In the forest, Michot reports hearing two and possibly three other woodpeckers drum and call. Michot was able to take a slide of one of the birds (which I have seen), but the view of the woodpecker was obstructed by leaves in the distance. The two lost sight of the bird as it flew into the forest. Michot returned to the same location on three occasions but never saw the bird again.

The experience left Michot shaking his head. "The bird had white on its trailing wing feathers, but it did not fly as straight or direct as ivory-bills supposedly fly. Also the call was a series of short notes, not the nasally 'yank, yank' that I have heard from an ivory-bill recording," he said. Over the years since his sighting, Michot has talked himself into believing it was an ivory-bill, only to question later whether it really could have been. When I first met Michot, before the Arkansas sighting, I sensed he almost wanted to persuade himself that he had not seen an ivory-bill. Today, while he will not say that he is 100 percent certain that the bird he saw was an ivory-bill, he is convinced they exist not only in Arkansas but in Louisiana as well. And within Louisiana, he believes the Atchafalaya Basin is probably the most likely spot.

Michot was not particularly secretive about his experience; instead, in 1985 he filed a written report with the U.S. Fish and Wildlife Service in response to their request for information pertaining to the ivory-bill. However, like many other folks who think they may have seen the bird, Michot never actively campaigned for the legitimacy of his sighting. Before the Arkansas sightings, Michot stated that he was hopeful the ivory-bill survived, and he thought that it did, but past destruction of the bottomland forests and the resulting habitat degradation concerned him. While he never gave up hope, he openly wondered whether even a few birds could have survived.

In the post-Arkansas era, Michot is now much more optimistic and excited about the potential for ivory-bills to be living in Louisiana and elsewhere. If they can survive in Arkansas, where many wrote them off years ago, surely the odds are good that a few birds have hung on in the Atchafalaya Basin and the Pearl River Wildlife Management Area, Michot said.

However, even with the Arkansas report, Michot is still uncertain about his own sighting. Some evidence seems to him to point in favor of an ivory-bill, while others factors do not. Michot debated the subject with me in an e-mail exchange in the fall of 2005. In response to my question as to whether or not the Arkansas report made him confident about his own sighting, Michot wrote:

> In some ways yes, some ways no. I was always confident that the bird was around, at least in the 1980s. There was a time period when I didn't hear about any sightings, until 1999 Kulivan. But now with all of the increased activity and buzz, it does make me feel better. I flew over the Duck Lake area again and it definitely has some big trees, which made me feel good. It is good habitat, and close to other sightings. All that is good. Our wing pattern was always the number one trait, and that still stands (white trailing feathers on the wings). I liked that several of the Arkansas sightings keyed in on exactly the same thing: white on trailing edge of wing. Both Hankla and I saw that and we saw it well. Everything else (size, flight pattern, call) does not seem to fit very well. Hankla saw a light "manila folder"–colored bill, which is good.
>
> My picture shows crest laid back, which is bad, but several paintings and pictures of ivory-bills show crest back, so that's not too bad.
>
> The call we heard was not ivory-bill-like, more like pileated, but not exactly. A series of single notes. We also heard drumming, but no double-rap that I have since heard so much about.
>
> Jackson says flight pattern is not a good trait. Our woodpecker had an undulating pattern. Audubon specifies that the IBWO had undulating pattern, which is good. Tanner (I think) said it does not, which is bad.

> The Luneau video did not look like our bird, which is bad. But, that was a different setting, so maybe ok. I am a little hesitant also about the size of our bird; it was big, but maybe not as big as a pintail.
>
> I keep going back to that wing pattern though. That is what really alerted us, and it is all we have, plus the bill color. Neither of us noticed the white on back of folded wings, so that is bad. But David Hankla was looking at the bill, and I was fooling with camera, so we really did not look. My picture shows red crest with black front, which is good, but could be either male ivory-bill or female pileated.
>
> So, in summary, I would say, yes, slightly more confident, but all in all, not much more. About the same. I guess I am about 50 percent sure that it is an ivory-bill.

This exchange makes obvious Michot's doubts about whether he saw an ivory-bill. And, given Michot's training as a scientist, the rarity of the bird, and the scrutiny to which reported sightings are subjected, such caution is understandable, even prudent. But Michot also describes himself as extremely optimistic and confident that the ivory-bill survives today, both in Louisiana and out of state as well. Michot again: "In Louisiana, I think that my optimism at this point in time is evenly split between the Pearl, upper Grand River, and the St. Mary Parish lower basin (around Franklin). Outside of Louisiana, I would say first is Arkansas, second is South Carolina (I have heard some buzz, but not many details), third is tie between Texas (Big Thicket) and Florida (several locations). I was pretty surprised when the bird was found in Arkansas; it was not on my radar screen. I had been focused on Louisiana. So, in a way, anything goes. The bird could be anywhere, and there might be a lot more out there than we think: maybe 25 or 50 individuals, maybe more!"

Wylie Barrow

Another member of Team Elvis South, Wylie Barrow is an ecologist with the USGS in Lafayette, like Tommy Michot. Barrow may be the member of the group most qualified to be termed an ivory-bill expert, although his research and writing is on avian ecology as well as bottomland hardwood forest ecology. Some of his research, including his dissertation work, took place in Tensas River National Wildlife Refuge, where Tanner conducted his study. Therefore, Barrow has had ivory-bills on his mind, in one way or another, for many years. In fact, Barrow was fortunate enough to meet with Tanner when Tanner returned to Tensas after it was designated a national wildlife refuge in 1983. Barrow said that while Tanner was impressed with the new reserve and how much the forest had matured since its destruction in the 1940s, his journey back to the Tensas River National Wildlife Refuge also had a bittersweet aspect, given that the birds had been long gone and that the forests in the reserve had been virtually destroyed.

When I met Barrow for the first time in 2003 near Lafayette, Louisiana, he had mixed feelings about the ivory-bill's chances of surviving. While the many reported sightings—such as the one reported by his friend and colleague Tommy Michot—gave him hope, the extent of human impacts on the ivory-bill's bottomland home made him skeptical. Since that first meeting, I have been fortunate enough to get to know Barrow better, and I know him to be a careful, even cautious biologist. His skepticism was thus based on hard scientific data—mainly the fact that so much habitat had been lost. However, even though he was cautious about speculating that the bird might still exist, Barrow really wanted to believe. After all, part of his dissertation research took place where the towering trees of the former Singer Tract had once stood. As a result of his research in the Tensas area, he knew not only Tanner's work but also some of the individuals who had assisted Tanner in the late 1930s. Before our first meeting ended, he gave me a copy of a photograph made from one of Tanner's original negatives stored at LSU.

Barrow's research background has allowed him to play a more prominent role in the 2005–6 search in southern Louisiana near Patterson. Thus, Barrow says that in certain ways, he feels as if his interest in the ivory-bill has come full circle, from looking at old black-and-white photos to searching for the bird. Barrow is now convinced that the ivory-bill does exist. Given the bird's low numbers, however, he remains concerned about its recovery. Like many I interviewed for this work, Barrow expressed the idea that we have been given a second chance to protect the ivory-bill, and that we should do everything possible to save the bird and its habitat.

Garrie Landry

Still another member of Team Elvis South is Garrie Landry, a botany instructor and the herbarium curator at the University of Louisiana at

Lafayette. Landry was among the first individuals I met during my research for this book, and he has proved incredibly helpful in my efforts to track down people connected to the ivory-bill's story. Landry, who lives in Franklin, helped arrange my initial contact with the outdoorsman Fielding Lewis. At that time I was aware of the photographs of an ivory-bill that Fielding had given to George Lowery back in 1971, but I did not know how to track down the photographer, whose identity Lowery had concealed. Landry pointed out that Fielding Lewis had actually written and self-published a book, *Tales of a Louisiana Duck Hunter,* in which he discussed the entire incident. So the story behind the photos was not as mysterious as I and many other people had thought.

Landry is personally fascinated with rare and extinct species. His house, for example, is in some ways a shrine to the passenger pigeon, with the living room filled with everything from past ads featuring the pigeon to artistic depictions of the bird and even an actual mounted specimen. Landry's interest in the ivory-bill, while it may not rival his fascination with the passenger pigeon, is intense, and this is hardly surprising, given his proximity—both professional and geographic—to the ivory-bill controversy. He knew George Lowery personally while a graduate student at LSU, and the mystery photographer, Fielding Lewis, lives a couple of miles from his front door. And today, Landry lives a short distance from some of the best and most recent reported sightings of ivory-bills.

When I first spoke with Landry in 2003, his viewpoint was similar to that of many other avid birders who had followed the ivory-bill story over the years. He believed the Fielding Lewis photos were authentic, and he thought the 1999 Kulivan sighting was sound, but the fact that the ivory-bill's habitat had been so altered by humans made him question its chances of survival. Yet he remained cautiously optimistic. As he said in that first meeting, there had been enough good reports that "you have to think twice about them. They can't all be mistaken."

In the post-Arkansas phase of the ivory-bill saga, Landry is "totally optimistic and downright certain" that they exist elsewhere, the Franklin area being the most logical area given the history of reports. "They are here, they have always been here," Landry wrote during our e-mail exchanges in the fall of 2005 and 2006. "It's now just a matter of time before we confirm it with multiple people seeing the birds. Louisiana has too much habitat that is still in many ways wilderness and unexplored, perfect places for a big, elusive bird to live and to remain unseen." Landry seconds the viewpoint of the LSU ornithologist Van Remsen that the ivory-bill has likely changed its behavior over the years due to negative interactions with humans over the past century. He believes that the bird Audubon reported as being loud and easily detected no longer exists. As the Arkansas reports documented, ivory-bills today have become extremely wary, the result of hundreds of years of persecution.

When asked, "What do you see as the greatest challenge now for a potential recovery of the species, or is that even possible?" Landry responded a bit differently from Chuck Hunter, whose job it is to help design the recovery effort: "We must first realize that this species has managed to survive all these decades and apparently to reproduce without any intervention from man. It's going to be rather hard to implement a recovery plan for a bird that by and large we cannot locate, cannot monitor, cannot even begin to count, but yet we know it's out there. It would seem the most practical course would be to preserve and protect the areas we suspect it is in from logging and development. Fortunately for this bird, it seems to be holding its own in remote areas that at this point don't really interest developers and don't get much human traffic. The only threat as I see it to the continued survival of the species is one of habitat preservation, particularly from the logging industry. As our cypress forest recovers from the total harvesting of the early 1900s, it's going to be important to set large areas aside as natural areas not to be disturbed. If we can do this, the IBW will be here for a long time. Just how long will depend on how much continuous area we do manage to protect."

While habitat protection makes good sense, it could end up being the most difficult task given that some land will be privately owned. While many landowners may gladly cooperate, some rural landowners in the Deep South are highly suspicious of the federal government and federal regulations. While this may seem a broad stereotype, I have had many conversations in the past couple of years with people who fear losing control of their land to the government and who would fiercely resist any such attempt. The area in the southern Atchafalaya Basin south of Louisiana State Highway 90 that holds so much promise for finding more ivory-bills is largely owned by private individuals and natural resource corporations, as is much of the Atchafalaya Basin. The interests of landowners in this area, therefore, will need to be balanced against the concerns and requirements of those charged with future ivory-bill management and the implementation of related conservation policies.

Dwight LeBlanc

The most recent addition to Team Elvis South is Dwight LeBlanc. LeBlanc is the Louisiana state director of the U.S. Department of Agriculture's Wildlife Services. Although his title may not indicate it, LeBlanc is in charge of dealing with "problem" animals. When people have a bear eating their trash or sugarcane, or when people think they see a large cat, they call LeBlanc. Needless to say, he has interesting stories to tell about human-animal encounters in Louisiana. Like the others on Team Elvis South, LeBlanc has had years of experience both as a wildlife professional and as a private citizen who spends time in the Louisiana outdoors. LeBlanc has keen insights into a variety of conservation issues, including endangered species such as the ivory-bill. And LeBlanc has a longer history of dealing with the ivory-bill than most. For example, LSU professor Bob Hamilton asked LeBlanc to search for the ivory-bill in the Grand River area following Hamilton's experience in that area in 1978 (when he heard the double-rap of a large woodpecker). And because he is deeply involved in the Louisiana natural resources community, LeBlanc has heard some solid reports of ivory-bills in recent years. These experiences have led him to follow the issue closely.

Thus, when I asked LeBlanc, in a 2005–6 e-mail exchange, if he was optimistic about the ivory-bill's survival and the possibility of finding the bird in Louisiana, he was quite positive: "I have held on (maybe subconsciously, until recently) to the belief that they [ivory-bills] were still around somewhere, but, I must say that the 1980s observations of the bird bolster the hope of finding one down here, as have all of the recent sightings that have been shared with and among the A-team. You know, there are a lot of skeptics out there, but dealing with the public, I don't believe that everyone out there is crazy. And, I don't think that biologists know all that there is to know. Another example is the number of cougar sightings in Louisiana—while most can be discounted, they are like ivory-bill sightings—people are seeing something but we don't know what it is yet; but time will prove or disprove what the sightings were."

Because LeBlanc works directly with the public on a daily basis, his primary concern for the long-term recovery of the bird is the potential obstacle posed by the resistance of private landowners to what they see as encroachments on their property rights. And since, as noted above, the vast majority of the forested wetlands in Louisiana are owned by private individuals, this is indeed a serious issue. LeBlanc works directly with disgruntled landowners who have complaints about destructive animals. He fears a backlash could take place if a bird is found on an uncooperative landowner's property. And, based on the recent availability of "kill the ivory-billed woodpecker" shirts and coffee mugs on the Internet, one must wonder if a backlash has not already begun. LeBlanc works on the front lines of the interface between conservation interests and private landowners' rights, and he is well aware of the precarious balance that must be maintained between the needs of animals (some of which, like bears, are potentially destructive) and the needs of people who are primarily concerned with their property rights (in terms of either destruction to their property caused by animals or, as in the case of ivory-bills, having their activities curtailed because of the presence of a rare animal). LeBlanc writes:

> The greatest challenge facing an ivory-bill recovery will be convincing landowners that the birds are an asset not a liability. Remember that we are a Southern state and a product of the "War of Northern Aggression" and the federal government is not to be trusted. I truly believe that a battle between private landowners who live off of their land surface (forestry, farming) and "environmentalists" will occur if the bird is found. This will be the challenge. It will take money to preserve habitat (may be considered a land grab by many), money to complete studies (a waste for some who think we have more pressing problems in the world), and money to prevent destruction of identified woodlands and birds (which is not in endless supply). You know, bears are still routinely poached by some folks despite their protection and stiff fines for illegal take. And who will be there to see one being killed or a nest tree being felled?—we won't be able to watch continuously, ever. And while lawsuits are being argued in the courts, there are some who will use the opportunity to make life a living hell for any remaining birds that may need vast

expanses of timber during the course of a year. Keep your eyes open for the pending lawsuit on designating critical habitat for bears—this will be a minor event compared to ivory-bill recovery issues.

While the above scenario concerns LeBlanc, he remains optimistic overall. Part of his optimism and excitement regarding the search for the ivory-bill is the mystery that nature could persist in the face of what humans have done to the southern environment over the past couple of centuries. Also, as LeBlanc said, he does not believe biologists know everything. For LeBlanc, the world is more interesting while some mystery remains in nature, although he would sure like to see an ivory-bill in Louisiana. That is one mystery he would like to help solve.

Nancy Higginbotham

Nancy Higginbotham, a biology instructor at the Southeastern Louisiana University, had an ivory-experience that provides an example of exactly what others have said they feared would happen to them if they reporting seeing the bird. Her former colleagues at the Louisiana Department of Wildlife and Fisheries endlessly teased her about her ivory-bill sighting. Higginbotham actually saw the bird twice—a male on the edge of the Pearl River forest in 1986 and a female while she was deer hunting in the Pearl River in 1987. Before the Arkansas sighting, she often regretted telling anyone she had seen the bird. In fact, when I finally tracked her down in 2003, she was willing to talk with me, but she also said, "I was afraid that you would track me down sooner or later." She wasn't being unfriendly; she was just tired of being hassled about her experiences. The 1999 Pearl River sighting did little to reduce the tension she felt about the issue because the personal scrutiny focused on David Kulivan as a result of his reported sighting was reminiscent of her own experiences. According to Higginbotham, very few people believed her story. One exception was Professor Bob Hamilton at LSU, with whom she took a class. In our conversation, Higginbotham fondly recalled Hamilton and his encouraging words about her experience.

In 1986, Higginbotham's family owned a fried-chicken restaurant on LA 41 on the edge of the Pearl River swamp. One afternoon, while she was sitting outside of the restaurant with her mother, a large, crested woodpecker landed on a tree in the back of their property. She and her mother immediately realized that they had never seen this bird before. Higginbotham had seen pileated woodpeckers, but this bird was much larger and more stunning. She turned to a birding guidebook and realized what she had seen was an ivory-billed woodpecker. When she told her father and others, they told her she was crazy. However, she had grown up spending time in the outdoors, she knew her birds, and she knew she was not crazy. She knew the bird was not a pileated. In her own words, from our 2003 conversation:

> It landed on a big oak tree out in front of the restaurant [Cluckers on Hwy 41]. It stayed on the side of the tree for about 2 minutes (not feeding, just hopping) and then flew away never to be seen again (by me anyway). It was close to us, within 50 feet. It was the biggest woodpecker I had ever seen, like 24 inches tall. It was as big as a chicken on the side of that tree and all of the markings were visible and the bill was obvious. We were shocked by the size and coloring of the bird—that was our (mine and my mom's) first thought. My Dad walked out and it flew off. When we told him what we had seen, he said it sounded like an ivory-bill but there was no way because they were extinct. After that I got out the bird book and researched the ivory-bill. I wrote that sighting off because I was uneducated as to the differences, but I still think that it was an ivory-bill.

A year after this sighting, in 1987, Higginbotham was deer hunting in the Pearl River Wildlife Management Area. She was quietly sitting in a deer stand when a female ivory-bill appeared and landed on an old, dying sweet gum. Higginbotham claims she was able to watch the bird most of the day because she did not move from the deer stand, and the bird kept returning to the dying tree, "entering different nest or roost holes." According to Higginbotham, the bird would leave, come back, occasionally call, enter the nest holes, and then fly off again. She had many hours in the deer stand to observe the bird. She is convinced that it was a female ivory-billed woodpecker.

As a result of her own experiences in the Pearl area, Higginbotham was not really surprised when she heard about the 1999 sighting in the same general location where she had been deer hunting. When the 2002 search ensued, she was employed as a nongame biologist at the Louisiana Natural Heritage Program, which is part of the Louisiana Department of Wildlife and Fisheries. In that capacity, she played a critical role in planning the search for the ivory-bill in Pearl River Wildlife Management Area for the State of Louisiana.

While Higginbotham was happy that a formal search for the bird took place in 2002, she was adamant that not enough territory had been covered by either the group or the audio recorders that were placed in the Pearl area after the team left. When we met in her office in 2003, she showed me a map and the locations where the recorders had been placed. I will admit I was surprised at the limited geographical area that they covered. Given the size of the area searched, Higginbotham said she was not surprised that a bird was neither seen nor heard. The fact that an ivory-bill was not found did not diminish her belief either in her own sightings or in bird's presence in the Pearl region. As she reiterated, "They [the search and recording devices] didn't cover half of the territory in Pearl. Those birds can cover a lot of ground. They are long-distance travelers, probably using different parts of Pearl and the other wilderness areas at different times of the year." Perhaps these birds and others use much of the Lower Mississippi Valley at different times of the year, traveling along the levee or batture forests between the larger sections of forest such as the Atchafalaya Basin or Tensas. The Mississippi River may be literally a large corridor for a small population of ivory-bills.

It was an emotional moment for Higginbotham when she heard the news about the Arkansas ivory-bill sighting. "When I heard the news, I cried for a week," she said. When asked if they were tears of joy, she expressed mixed feelings: "I was excited about them finally finding the bird, but I was also really upset that I did not play any role in it, and won't play any role in more searches in Louisiana." Higginbotham said she would not have left the Louisiana Department of Wildlife and Fisheries if she thought they might really find the bird during an organized search.

Since the Arkansas report, Higginbotham has taken some pleasure in teasing the very people who have hassled her over the years about her ivory-bill sightings. The Arkansas report has assured her that the ivory-bill exists in the Pearl area as well as the Atchafalaya Basin.

"I know they are still here," she said.

Post-Arkansas

While the experts interviewed for this book are thrilled to varying degrees about the ivory-bill's rediscovery, they all agree that even if the ivory-bill is found in other locations it faces great challenges regarding its recovery. All also emphasized that forestry-management practices must be altered to create both more diverse bottomland forests containing species favored by the ivory-bill and more climaxlike forests. They also agreed that it appears we know very little about the present-day ivory-bill's ecology and behavior. For example, based on Tanner's work, people assumed the ivory-bill was a raucous, gregarious bird that needed extensive tracts of old-growth forest. The ivory-bill seems to be much more adaptable in its use of habitat than was previously suspected. But while the experts are awed at the ivory-bill's ability to hang on, they are also cognizant of the biological realities of small populations, limited habitat, and minimal ecological information about the present-day ivory-bill versus the ivory-bills of the 1930s.

4

LOCAL STORIES

I was not born and raised in Louisiana, nor do I live in a rural locale. Without the help of the local people who took the time to meet with me and to allow me to briefly enter their worlds, I could never have learned to see the landscape as they do. When I was out in the forest with the people who still live near it and use its resources, I came to realize how different my world is from theirs. Though I may have lived only a short distance from some of the towns bordering the Atchafalaya Basin, my daily life might as well have been a thousand miles away.

The differences between our worlds were exemplified by an incident that took place while I was interviewing some folks near Bayou Pigeon in the Atchafalaya Basin. I had a bumper sticker on my car from the Coastal Conservation Association, or CCA, a national conservation organization focused on the management of game fish and coastal environments. These bumper stickers are not uncommon around Baton Rouge. Many sport fishermen in the Gulf Coast region have the distinctive redfish bumper sticker on their cars, especially in cities, where many more sport than commercial fishermen live. The CCA is not a radical environmental group by any means. I was a bit taken aback, therefore, when a couple of commercial fishermen with whom I was speaking began to harass me—in a good-natured, but serious way—about my bumper sticker and my ultimate intentions regarding the ivory-bill search. "Since you're an environmentalist, are you here to find the bird and confiscate our land?" one asked. Before researching this book, I never would have imagined such a reaction to the CCA. Yet, in some parts of Louisiana—and probably other states that are home to significant numbers of commercial fishermen—the CCA is seen as the enemy because it has fought to ban activities such as gill netting. And given the collapse of the commercial fishing economy in Louisiana, suspicion and anger toward groups that have curtailed some of their activities are understandable.

This experience forcefully brought home to me that outside the comfortable confines of Baton Rouge, many individuals see the environment and environmentalists far differently than I do. This difference in perception had significant bearing on the responses of many of the local people I interviewed to the search for the ivory-bill and the efforts that may be undertaken to protect its habitat. The majestic bird that most bird-watchers see as a critically endangered species deserving of protection—such as that afforded by the Endangered Species Act—many rural people see as a threat to their land and livelihoods. And while I may not share their perspective, the conversations I have had with these folks while researching this book have given me a greater appreciation of their concerns about the intrusion of outsiders into their lives.

The people whose stories are collected here have an intimate knowledge of the bottomland forests where some ivory-bills may still live. Stories of past duck hunts and encounters with water moccasins, along with recollections of wolves and bears, were some of what they shared with me. And some of these people still remember when an ivory-bill sighting was not front-page news. One elderly man told me that in the 1930s, he regularly saw ivory-bills while deer hunting. He said that the popular name for the bird back then was the "Good God Bird," the exclamation that the sight of such a large and beautiful woodpecker was thought likely to evoke. A couple of other elderly people stated that they did indeed remember the ivory-bill, which they called the "forest turkey." "Back in the 1930s," one man stated, "when money was scarce, we used to eat it just like a turkey. You ain't picky when money and work is scarce." In both the Pearl River area and the Atchafalaya Basin, people told me that they had dined on the forest turkey. When asked what the forest turkey tasted like, one informant gave me a wry smile and said "chicken." Besides providing insight into how people struggled to survive in rural Louisiana during the Great Depression, stories like these illuminate the relationship that many of these people have with their environment.

The stories of local individuals need to be heard and recorded, not just because they are interesting, but because these are the people who know the ivory-bill habitat best. Many within the birding and ornithology communities have dismissed outright—as either lies or misidentifications of the pileated woodpecker—the many credible reports of ivory-bills that have come from individuals outside of the scientific establishment.

But should these sightings from the people who best know the bottomland forests be so quickly dismissed? These are reports from the people who spend the most time in this remote and often inaccessible environment, whose identities and livelihoods have long been rooted there, and who take pride in their knowledge of it. Like LSU professor Vernon Wright, I believe that if these sightings had been thoroughly investigated, the existence of the ivory-bill would not be a subject of debate today.

I did not believe that everyone who approached me with a woodpecker story had seen an ivory-bill. I knew that the accuracy of my account depended on approaching all reports with a healthy dose of skepticism. But it didn't take me very long to winnow out the mistaken ivory-bill identifications from those that were credible. Several individuals, for instance, reported seeing woodpeckers with white backs clinging to tree trunks. Yes, I told them, ivory-bills show white on their backs (trailing wing feathers) when their wings are closed. But the size of the birds they described made it quickly apparent that what they had seen were red-headed woodpeckers (*Melanerpes erythrocephalus*), which also have white backs when they close their wings. But if the informant was already familiar with the pileated, if he or she could clearly describe a specific ivory-bill feature, and if the bird was sighted in an area that could realistically support ivory-bills, I took the information more seriously.

While I have been shown few photos—and those that exist are not very clear—I have become certain over the course of my research for this book that several people have seen the ivory-billed woodpecker in Louisiana during the past thirty years. There have been time gaps between sightings. And, interestingly, some of the best reports have come from people who don't consider themselves birders—or they didn't at the time. The experience of sighting an ivory-bill has often piqued the interest of the observer, creating a new fascination not only with birds but with their habitats as well.

My conversations with rural folks have led me to believe that when a good photograph or video recording of an ivory-bill does surface, it probably will have been captured by a rural resident with little experience or even interest in bird-watching per se. In my view, the individual most likely to find the bird will be a hunter or a fisherman who spends a great deal of time in bottomland forests. These environments, as I have said, are not for the faint of heart. The mosquitoes, snakes, alligators, and prickly palmettos often prove powerful deterrents. In Louisiana, the Sportsman's Paradise, fishing and hunting are the two activities that draw most individuals into the bottomland forests, where the deer, ducks, and fish are found in the greatest numbers. The prevalence of hunting clubs, which seem to have sprung up everywhere that swamp, marsh, fish, and waterfowl exist in the rural South, especially in Louisiana, indicates the popularity of these activities. While outdoor activities such as birding and canoeing are becoming more popular, by far the largest number of people who enter ivory-bill habitat are hunters and fishermen. These people know the landscape well—in many ways it is their landscape. To discount their reported ivory-bill sightings is shortsighted.

Just as the number of ivory-bills itself has declined, so has the number of people who earn a living from and spend time in the bottomland forests and marshes, even in Sportsman's Paradise. In addition to the valuable information about the existence of the ivory-bill they provided, my conversations with rural folk brought to light the fact that their way of life is in decline, especially in places such as the Atchafalaya Basin. For example, the number of full-time commercial fishermen has dramatically decreased over the past two decades due to competition from farm-raised catfish and crawfish and foreign imports of cheap seafood. It was often impressed on me during these interviews that the lifeways of these rural people were vanishing along with the bottomland forest they know so well.

Collecting information from these local people presented some challenges. As illustrated by the encounter I described above with the commercial fishermen who were suspicious of my CCA bumper sticker, some people were initially suspicious or reluctant to talk with me because they suspected I was a "Fed" who might confiscate their land or their friend's land where they had seen the bird. One person who saw an ivory-bill on state-owned land even told me that he would consider shooting the bird if it showed up on his property. I am still not sure if this person was serious or just trying to intimidate an outsider, but I had to assure many people that I had no such intentions, and that they could remain anonymous if they desired. While rural folk in the South love the outdoors, many bristle at the thought of outsiders coming in and regulating what they do on their land. For some, my informants' anonymity may call into question the credibility of their stories. Had I not been willing to guarantee these people anonymity, however, I believe the important information about their ivory-bill sightings that is gathered here—information that may be critical to finding and protecting the remaining ivory-bills—would not have been available.

The stories collected here call attention to an important question: Was the ivory-bill really "rediscovered" in Arkansas? Who determines what and when something is rediscovered? Did Fielding Lewis, Jay Boe, Scott Ramsey, Nancy Higginbotham, David Kulivan, and the many other individuals who reported seeing ivory-bills over the years think the ivory-bill was extinct until "rediscovered" by the Cornell and U.S. Fish and Wildlife team in Arkansas in 2004? How many opportunities to find the ivory-bill have been missed over the past several decades because the scientific and birding establishments discount credible information that comes from outside of their own domains?

The 2004 report of an ivory-bill that led to the Arkansas search were made by a kayaker, while various other reports from around the South, some highlighted herein, were made by outdoorspeople whose professions or recreation take them deep into ivory-bill habitat on a regular basis. Another important question—raised by wildlife experts and non-experts alike—surfaces often in this work: *Prior to the 2004 Arkansas search, how much time was spent searching for the ivory-bill, and how much territory was actually covered?* Canoeing, hiking levee roads, and even flying over forests are useful aspects of the search, but they are not comprehensive enough to guarantee that a secretive bird will be found in this often inaccessible environment. Given that vast areas of the potential ivory-bill habitat remain unexplored, the fact that these important, but clearly limited, searches haven't resulted in an undisputed photograph of an ivory-bill does not offer a sound, scientific basis for an argument that ivory-bills do not exist.

The Mystery Photographs: Fielding Lewis and George Lowery

There is perhaps no greater controversy surrounding the post-Tanner ivory-bill sightings than the debate occasioned by the "mystery photographs" of an ivory-bill taken in 1971 in Louisiana by outdoorsman Fielding Lewis. While the photos are more than thirty years old, they continue to be the subject of discussion in the birding and ornithological worlds. While it may seem strange to those outside the scientific and birding communities that a controversy over a few old photographs could be kept simmering for thirty years, the very fact that the photos have remained so controversial demonstrates the intensity of emotions associated with the search for the ivory-billed woodpecker. The photos became a lightning rod for debate not only because they were of a supposedly extinct bird, but also because the name of the photographer and the exact location in which the photos were shot were kept secret—at least from the general birding public—until fairly recently. Fielding Lewis took the photos near LA 317, adjacent to the Intracoastal Waterway southeast of Franklin. However, the story of these photos and the location in which they were taken cannot be told without including the role of Professor George Lowery Jr. Then director of the Museum of Natural Science at LSU, Lowery was the individual who presented the photos to the ornithological community, a decision that would affect the rest of his life.

The photos that Lowery presented to the ornithological establishment were met with great skepticism, including charges that the photographs showed a stuffed bird placed high in a tree in the middle of the swamp. Not only the photographer but also Lowery have been accused of faking the photographs. At that time the photos were released, Lowery was already a nationally known and respected ornithologist and the preeminent authority on Louisiana birds. He himself had seen some of the last known ivory-bills in the Singer Tract in 1933. While skepticism regarding the accuracy of the identification of an extremely rare bird is appropriate, the charges of fraud seem extreme and far-fetched, given Lowery's reputation. And what would motivate the photographer—a nonbirder who wished to remain anonymous—to perpetrate such a deception? Without a stake in the ivory-bill controversy or a desire for publicity, the photographer would seem to have little motivation to carry out such an act.

While some individuals continue to swear the photos are fakes, others, such as Van Remsen at LSU, are convinced they were authentic. In fact, Remsen told me, during our first meeting in 2003, that: "I would not bet my life the photos are real, but I would not bet my life on anything. I would bet anything else though, that the photos are authentic." In telephone conversations with Lewis in the late 1980s, Remsen became convinced that Lewis had seen ivory-bill woodpeckers on several occasions. But even with experts like Lowery and Remsen expressing their belief in the authenticity of Lewis's photos, some individuals continue to insist that the bird in the photographs is a mounted specimen placed high in a tree.

The photos were taken with an instant camera and therefore are not of the best quality. But the rather blurry photos cannot conceal that the bird in the photos is unmistakably an ivory-billed woodpecker. The white back feathers—one of the features that most clearly distinguishes the ivory-bill from the pileated—are clearly visible on the large, crested woodpecker. In fact, I am unaware of anyone who disputes that the bird in the photos is an ivory-bill; the point of contention is simply whether the photos show a stuffed specimen.

When Lowery received the photos in early 1971, he was surprised that the bird in the picture was an ivory-bill, because he, like Tanner, thought that the Singer Tract had been the ivory-bill's last refuge, and also because so many people confused the ivory-bill with the pileated woodpecker. In fact, in his *Birds of Louisiana,* first published in 1955, Lowery laments that, "It is possible that no future generation of Americans will be able to spend Christmas morning [the day when he saw the birds], or any morning, watching four Ivory-billed Woodpeckers go about their daily routine amid huge red gums whose diameters are greater than the distance a man can stretch his arms. I wonder what beauties we shall have, aside from the mountains and the sky, a hundred years from now!" (Lowery 1974:416). In the second edition of the same book, Lowery vaguely describes the location and the story of the photographs.

In all of the conversations I had while researching this book, I met no more interesting person than Fielding Lewis. The classic southern sportsman, Lewis exhibited a dedication to hunting and fishing of a sort I have rarely encountered. He is also a well-connected political and economic figure in Louisiana, especially in south Louisiana. Lewis is not a birder; he is a hunter, a fisherman, and an avid boxing fan who has served as chairman of the Louisiana Boxing Commission. According to Lewis, when they were both high school students in Franklin, he twice knocked down two-term governor Mike Foster.

Franklin, Louisiana, where Lewis was raised, is an ideal location for a sportsman. The town sits between the southern edge of the Atchafalaya

Basin, located to the northeast, and the coastal swamp and marsh to the south. To say the area offers a diversity of fish and game seems an understatement. Wildlife found within a stone's throw of Franklin include huge alligators, black bears, some say red wolves (their continued existence is a topic of debate), millions of migrating birds and waterfowl, alligator snapping turtles that can bite a paddle in half, and alligator gar the length of a small car. In his Franklin office, Lewis once showed me the mounted head of an alligator he had killed. I have seen many large alligators while canoeing, but this behemoth was by far the largest. When it was still alive and tormenting nutria, it was more than fourteen feet long.

Lewis was born in 1930; he moved to Franklin in 1935 and was raised there by his grandmother. When he was a young boy, some of the older hunters, trappers, and fishermen in town took an interest in him and brought him along on their outings. Lewis has many fascinating stories of his adventures in the swamps and marshes, as well as stories about some of the folk who lived off the local fish and game. Curiously, despite his long history in the outdoors around Franklin, Lewis did not see his first ivory-bill until the late 1960s. Perhaps the birds were extremely reclusive, and he simply never crossed paths with them. Another explanation is that the birds may have moved into the area in the late 1960s, as migrants from the heart of the Atchafalaya Basin, located to the north of Franklin. Once Lewis saw his first bird, though, he did see them on several occasions in the same general area until 1988. According to Lewis, when he discussed his sightings with friends around Franklin, others who had spent time in the area hunting and fishing also claimed to have seen the ivory-bill. In September 2003, I was able to speak with one of these individuals, Wilbert Cole, who claimed to have regularly seen the woodpecker in the same swamp until 1995, when poor health forced him to give up hunting and fishing.

The first ivory-bill that Lewis saw in the late 1960s was a male that flew across a road in front of him as he was driving on LA 317. Although Lewis wasn't sure then how to identify the bird he saw, he knew it wasn't a pileated woodpecker because of its size and the large amount of white on the trailing section of the wings. When he got home, Lewis looked in his bird book to try to identify what he had seen. And sure enough, as soon as he turned to the page containing the ivory-bill, he knew that was the bird that had crossed his path.

Lewis read in his field guide that the last ivory-bill had been seen in northern Louisiana in the 1940s. While he had immediately recognized that this was a rare bird simply because he had never seen it before, he did not realize until reading the field guide that the bird was thought to be extinct. Lewis described the bird he had seen to the Franklin Senior High School science teacher, who directed him to George Lowery at LSU. Although Lowery was a well-known professor, Lewis, who had little previous interest in ornithology, did not know him.

Lowery had received many ivory-bill reports over the years, but this one provided enough detail that he invited Lewis to stop by his museum the next time he was in Baton Rouge. According to Lewis, when the men first met, Lowery greeted his report with some skepticism. Even though Lewis had emphasized that he was familiar with the pileated and that the bird he had seen was different, Lowery carefully explored Lewis's understanding of the differences between the ivory-bill and the pileated. While Lowery was disappointed that Lewis had not seen the bird's white bill—in Lowery's view, one of the most outstanding features of the bird—Lewis provided other details such as the white on the wings and a flight pattern that was straight, rather than undulating like that of the pileated, that made Lowery think Lewis might have seen an ivory-bill. He maintained his somewhat skeptical stance, however, telling Lewis that "most sightings of ivory-bills turn out to be pileateds." When he left the museum, Lewis promised Lowery he would keep an eye and ear open for the ivory-bill.

In the fall of 1970, after duck hunting with a friend approximately three miles from the location of the original sighting, Lewis saw a pair of ivory-bills in an opening in the swamp. This time, he did see the bills and heard the "yamp, yamp" call that Lowery had described the year before. One outstanding feature of this event was that Lewis clearly saw a black-crested female. For a moment, Lewis thought about bringing one of the birds down with his shotgun, reasoning that if Lowery wanted proof, he would give him proof. However, he thought better of that action, and simply admired the birds as they flew deeper into the swamp and out of sight.

The next year, 1971, Lewis again saw ivory-bills while training one of his retrievers on the edge of the swamp near the first location where he had seen the birds. A pair of ivory-bills flew past him and into the swamp opening. He quickly put away his dog, grabbed his camera, and quietly followed the path of the birds. The male had landed on the trunk of a tree in front of Lewis, allowing Lewis to start taking photographs of the bird with his instant camera. He was able to get within thirty or so feet of the male bird. Lewis lost sight of the female bird, but he had an unobstructed view of the male for several minutes. It was during this sighting that Lewis took the photos deemed so controversial by many in the birding and ornithological communities.

Lewis promptly sent a couple of the photos to Lowery at LSU. Lowery's review of these photos convinced him that Lewis had indeed photographed an ivory-bill. Lowery soon made arrangements to venture with Lewis into the bottomland forest. Lowery decided not to reveal the location where the photos were taken or the name of the photographer because he did not want the birding community to overrun the area. Lewis said he thought at the time that keeping the location secret was rather strange; at that time he had no idea that many birders might travel long distances to see the ivory-bill.

Lowery never did reveal Lewis's identity or the location of the bird. According to former students and Fielding Lewis, Lowery anticipated the criticism for this secrecy that was later leveled at him by many in the birding community. But that did not deter him from trying to protect what could be a few pairs of ivory-bills in the area from being disturbed by the throngs of birders who would be likely to invade their swamp forest. And Lewis told me that he himself had had reservations about "going public" because he had taken the photos on private land and he did not want federal or state authorizes to intervene and seize the land.

In a later edition of his *Birds of Louisiana,* Lowery offered only the general information that the photographs were taken south of LA 90. This further confused and exasperated some critics, who thought that Lowery was intentionally misleading them. Some birders assumed that the photographs were probably taken south of U.S. 190, in the Atchafalaya Basin. These criticisms and accusations of fraud remained a sore point for Lowery until his death in 1978 (Nevin 1974; Gallagher 2005). Regarding the controversy surrounding the photos, Lowery stated: "I just wish I had seen it [the ivory-bill] myself . . . and then I wouldn't give a damn how many people questioned it" (Nevin 1974: 80).

While there is some bottomland forest south of LA 90 (much of it dominated by cypress), it is not nearly as extensive as that found in the Atchafalaya Basin. Much of the forest there is more open than that located in the Atchafalaya Basin because the area is so near the coast and salt water. Generally speaking, the closer the cypress stand is to a saltwater coast, the more these saltwater-intolerant trees tend to be dispersed. Lewis, like many others, doubts that the Atchafalaya Basin can support ivory-bills because, in his words, it is full of "trash trees, and not the big cypress found around here."

Lowery and his wife visited the Franklin area shortly after the 1971 sighting. Although they did not see the ivory-bill on that trip or on any subsequent trip to the Franklin area, they did find signs of the presence of large woodpeckers. For example, a large tree in the exact location where Lewis had originally spotted the pair of birds was in the process of being excavated by large woodpeckers. Lewis was confident that the hole had been created by ivory-bills since he had seen an ivory-bill in the vicinity and, he believed, even on the very same tree, but, sadly, when he returned a couple of months later, it appeared to have been abandoned. On another visit, when they saw a pileated woodpecker fly across their path, Lowery immediately shot a skeptical look at Lewis and asked, "Is that the bird you saw?" Lewis immediately fired back, "That was a pileated, and that is not the bird I told you about!" Toward the end of the 1970s, Lowery's declining health kept him from going into the field to search for the bird. Even though he never saw the ivory-bill with Fielding Lewis, he remained confident that Lewis had indeed seen the bird.

After his initial encounter with the ivory-bill in the late 1960s and his subsequent interactions with Lowery, Lewis began mapping the locations of not only his sightings but also those of others he considered reliable. Lewis eagerly points out the area where he believes ivory-bills existed in the past and continue to exist today. He is confident that they are still present in this area because it remains intact, with little alteration since the time he first began seeing the ivory-bill. According to Lewis, this relatively small zone of old-growth cypress remains the wildest place in Louisiana.

Today, Fielding Lewis doesn't get out into the swamp very often. He has given up his favorite pastime of duck hunting. He is also much less interested in keeping the location of the ivory-bill a secret. He was not at all reluctant to talk with me and to provide specific details about his sightings. While he remains suspicious of the motives underlying state and federal conservation initiatives, he says he truly wants to protect the ivory-bill. In fact, he told me that he actually had contacted the Nature Conservancy about possibly purchasing the old-growth cypress swamp. I asked Lewis if he would be interested in taking me into the swamp in the area where he had previously seen ivory-bills. While Lewis struggles with an injured knee—an old sports injury—and diabetes, he responded enthusiastically to the idea of showing me his old hunting grounds.

My anticipation of visiting the swamp with someone I am convinced saw and photographed ivory-bills made the drive to Franklin exciting for me. And that I was going into the swamp with someone who knew the area intimately was reassuring after some of my previous adventures. Accompanying us was John Richard, an old hunting friend of Lewis's with whom he had seen an ivory-bill fly across Highway 317 in the late 1960s. Richard and Lewis have shared many hunting and fishing experiences over the years, and spending the day with them made me feel as if I were living in an issue of *Field and Stream*.

We put Richard's boat in at an oil-field exploration landing. Huge decrepit engines sat rusting on the banks of an access canal. This was not exactly a pristine birding landscape, I thought. But then again, who knows what might be flying around in here? Few, if any, birders or ornithologists actually come out here or can even get permission to be on property like this. We were traveling by boat because there is little dry land in this region of the state. Although there are spoil bank levees that crisscross the area, it is inundated with water.

This part of Louisiana is a transition area between wooded swamp to the north in the Atchafalaya Basin and grassy marsh to the south. Although you cannot smell the ocean or see any beach, this area is only a few miles from the Gulf of Mexico. This environment was unlike any other I had previously visited. It was not densely forested like the Atchafalaya Basin just a few miles away. The ground cover was dominated by marsh grasses, and while the "forest" was open, there were many very large cypress trees, most having dead snags with woodpecker holes. In fact, I was struck by just how extensive the woodpecker "damage" was. There were also many large nest holes. Most of the damage can be attributed to pileated woodpeckers, and, in fact, we saw and heard several.

However, given the number of woodpecker excavations, dead snags, and large trees, it is conceivable that ivory-bills continue to inhabit the area.

It is also very likely that a small number of ivory-bills could live in this area without being detected. The ground cover was inundated with waist-deep water and choked with grasses, making almost impossible any attempts to slog through in search of the birds. Even traveling by boat is difficult because many of the canals are shallow and overgrown with water hyacinth. The entire area, with its tropical feel and look, reminded me of a scene from *The African Queen* in which Humphrey Bogart was forced to drag his boat through the weeds and mud. The only people who spend any time in this area are oil-field workers on large portable drilling barges and the occasional hunter and fisherman.

I left the area feeling overwhelmed. It wasn't the first time, in the course of researching this book, that I had reacted this way to a visit to one of the many large, wild spaces in southern Louisiana. How in the world, I wondered, can we definitively state the ivory-bill no longer exists in one or all of these remote areas? It's true that these landscapes aren't pristine, and in some cases they are severely degraded and even polluted by activities such as oil drilling. However, many areas still remain wild and largely unexplored, at least by birders, ornithologists, and other naturalists. And while the area we had just left was not pristine, Calumet Swamp nearby is a large cypress swamp containing virgin timber. Though it is not a huge area, Calumet could have provided—and could continue to provide—a refuge for ivory-bills, And given the presence today of ivory-bills in this area suggested by preliminary reports from the 2005–6 search, that is exactly what appears to have happened.

In support of the belief that the ivory-bill still survives in this area, Fielding Lewis points out—referring specifically to logging—that nothing has changed about the local environment. So, in his mind, no human actions would have driven the birds away. Certainly one could argue that a small population here might be doomed either by inbreeding or by a catastrophic event like a hurricane, which could easily wipe out an entire population. Or perhaps the ivory-bills Lewis spotted were remnant individuals that have since died of old age. But given that the habitat has changed very little and that there have been credible sightings in the area as recently as the summer of 2005, I understand why Lewis continues to believe the birds are in that swamp.

At first, I thought it rather ironic that Lewis, of all people, had spotted and identified the ivory-billed woodpecker. First, he is not a birder, and second, he is a dedicated activist for landowners' rights. He is deeply suspicious of efforts by both state and federal government to dictate what rural people do with their land and the resources on that land. Although I knew that many landowners' rights activists appreciate the outdoors or wildlife, this was not the type of person I once thought was likely to sight a very rare bird. After getting to know Lewis, however, I came to realize that his skepticism about the methods and motives of conservationists existed alongside a deep affection for and knowledge of the forests and marsh of south Louisiana and the wildlife that inhabits them. I believe that Lewis has come to feel a profound sense of pride that his backyard, so to speak, was—and likely still is—home to the ivory-billed woodpecker.

I would probably be wrong to say that Fielding Lewis feels vindicated after the sighting of the Arkansas ivory-bill since he never once doubted what he saw in the 1960s, 1970s, and 1980s. It would more accurate to say that Lewis seems happy that someone else has finally found the bird that he never questioned was still flying. During our conversations, he seemed unconcerned that outsiders doubted his photos. When it comes to wildlife in the place he knows so well, he puts little faith in the opinions of city slickers.

Patterson, Louisiana

Since Fielding Lewis first drew the birding world's attention to the far southern edge of the Atchafalaya Basin with his ivory-bill photos and his relationship with George Lowery in the 1970s, reports of ivory-bills from this region of Louisiana have continued to surface. For years, locals who hunt and fish in the area's forests south of U.S. 90 have claimed to have seen large black-and-white woodpeckers in the surrounding bottomland forests and marsh. In fact, given the credibility of the reported sightings in this area and the quality of its habitat, I predict that a documented ivory-bill sighting or sightings in Louisiana will come from the area between Weeks and Morgan City (map 1), and that it will be the local people who spend time in the swamps who will lead the experts to the bird. And given that state and federal authorities as well as the Nature Conservancy have recently searched this area (and reported sightings), the filming or photographing of an ivory-bill will probably come sooner rather than later from this area that is off the beaten path for birders.

Scott Ramsey

An attorney and landowner who lives in Bayou Vista, Louisiana, Scott Ramsey has provided one of the credible reports from this area. Ramsey, like Fielding Lewis, is a classic Louisiana sportsman. He is also part of a new generation of sportsmen with an appreciation for the conservation and management of nongame species such as the ivory-billed woodpecker. Ramsey may not know the songbirds as well as the game birds, but he is pleased that his property, as he manages it, provides habitat for a variety of species. In fact, many of the weekend birders who distrust ivory-bill reports made by local people and sportsmen may have someone like Ramsey to thank if the birds are actually breeding in Louisiana. Hunting clubs and private landowners have protected the remaining tracts of bottomland forest. While their efforts have not necessarily been directed toward preserving the ivory-bill, the bird has benefited nonetheless.

Scott Ramsey was first brought to my attention by Fielding Lewis, who claimed he "knew a lawyer over in Patterson that has birds [ivory-bills] all over his property." Needless to say, I didn't really think Ramsey had ivory-bills all over his property. However, I now know that he does have at least a few ivory-bills on his land. I also learned through that experience never to doubt Fielding Lewis. Ramsey was not easy to approach. He has 18,000 acres of bottomland hardwoods and marsh that he uses as a private hunting area, and he doesn't want a bunch of birders tramping though it. Not only does he not want his property disturbed, but he also worries about outsiders on the land because it harbors some pretty dangerous residents such as wild pigs, bears, water moccasins, and canebrake rattlesnakes (*Crotalus horridus atricaudatus*).

Like many local people I have met while researching this book, Ramsey was generous enough to meet with me in July 2005 and show me around his property. There are hunting camps, and then there is Ramsey's camp. Just outside of Patterson, he has constructed a camp that is a hunter's dream. It consists of maturing bottomland forest, dissected by cleared paths; small fields planted with beans to attract deer; and a fine small house filled with trophies of past successful hunts. The property is contiguous with the vast bottomland and cypress swamps south of U.S. 90 that have produced so many solid ivory-bill reports over the years.

We toured Ramsey's property on a high-performance golf cart equipped with larger tires and a more powerful engine to get through the mud holes and over the fallen tree branches. Having slogged through similar forests on foot many times, I could not help but enjoy the comfortable ride, although driving through the spider webs, home to huge garden weaver spiders, made for an interesting experience. The forest contained many large trees, including some areas dominated by relatively young hardwood species. It was the classic bottomland hardwood forest—with leaves dripping water from the day's 99 percent humidity, Spanish moss, frogs chirping from the small bayous running through the area, warblers flitting along the path, and armadillos scurrying over the trail.

According to Ramsey, the forest had been damaged by past hurricanes and a forest fire, and some sections were clearly in a regenerative stage. However, like the region of Arkansas where the ivory-bill was found, it had enough mature trees, along with many damaged trees, to support a population of ivory-bills (as part of a larger resource region). Ramsey pointed out many rotting trees on the forest floor. These trees provide an ideal habitat for the grubs that make up a large percentage of the ivory-bill's diet. Louisiana black bears, like the ivory-bill, feast on the grubs living in the dead trees, and this region houses the largest number of Louisiana black bears in the southern part of the state. In fact, state wildlife officials have conducted bear surveys on Ramsey's land.

Ramsey made an interesting point about the bears, one that I had never heard from anyone else. He said he believes there is a connection between bears and ivory-bills. According to Ramsey, "bears hibernate in the old, hollow cypress trees. They also eat many of the same things—fruit, grubs, etcetera. So I believe where you find healthy bear populations, you might find ivory-bills." This observation makes sense. Bears and ivory-bills would use the old, hollow trees left by loggers (because they were hollow). Tensas River National Wildlife Refuge houses the largest population of bears in the northern half of the state, and as will be discussed later, there have been a few reported ivory-bill sightings from Tensas River National Wildlife Refuge in recent years.

Ramsey is not a birder per se, and until recently he paid little attention to the various woodpeckers on his land. He said he has seen ivory-bills off and on over the years, but it wasn't until the spring of 2004 that he set his sights on documenting the fact that he has ivory-bills on his property. Ramsey spends a great deal of time on his property, often traveling through the forest on his almost-silent golf cart. He also often sits in the various deer stands on his property, remaining silent and hidden from birds and other wildlife. His camp is close to his home, and Ramsey calls it his psychiatrist, a place where he can get away from the pressures of work. Again, it is individuals like Ramsey who spend a great deal of time in the ivory-bill's habitat who have the best chance of seeing the bird.

While Ramsey has seen ivory-bills on several occasions, these sightings have been brief. He concurs with the Arkansas research team about the bird's wariness. According to Ramsey, the bird doesn't sit still for very long and can only be approached by people if they remain silent. Ramsey commented several times that the bird appears out of nowhere, from high in or over the tops of the trees.

I asked Ramsey if he feared what might happen once the ivory-bill is indisputably documented to be on his land. In other words, did he have any worries about inviting state and federal officials onto his property to look for an endangered species? He said he had given the matter much thought. But, based on past positive experiences with conservation officials who have studied bears on his property, he feels confident that his property and hunting rights will remain safe. As he pointed out, he has no plans to log the area. He wants the land for hunting and has no need for minimal timber revenues. Ramsey recognizes the risks that a documented sighting of an ivory-bill could pose to his land rights. Unlike some other areas that may have ivory-bills, this area is all private property. Despite the potential risks, however, he feels connected enough to the land and to the wildlife that inhabits it that he wants to make sure the ivory-bill is protected, and even studied, on his property. He takes pride in the fact that he has managed his land in an ecologically sound enough manner that it houses a rare bird. Also, Ramsey believes that if the federal government either seized his land or limited his access to it, so much negative publicity would be generated that, at best, no one would report ivory-bills on their property, and at worst, they might even kill birds if given the opportunity. That, of course, is the worst-case scenario; Ramsey says he is confident that the goals of ivory-bill conservation and his hunting activities are compatible.

While we did not see an ivory-bill during our time in the woods, Ramsey showed me a roost tree where a bird had been seen one week prior to my visit in July 2005. The oval-shaped or oblong entrance hole was exactly as Tanner describes it. If one were to place a photo of the cavity hole next to Tanner's old photos, you could not tell the difference between them. As we waited in our swamp buggy near the roost tree, I worried that a bit of the ivory-bill bad luck—that there's almost never a camera handy when sightings of the bird are made—might have crept into the scene. I had my camera with me, and Ramsey had a video camera. As I tried to take pictures of the tree, my camera literally locked up. At the same time, his video camera would not record. I felt a bit panicked. An ivory-bill had been seen just a week prior to my visit, and now my camera would not allow me to take a picture! At that moment, I truly believed an ivory-bill was about to land on the tree.

When discussing where he has seen ivory-bills, Ramsey made an interesting point: he most often sees the bird on the forest edges. Perhaps the location of those sightings is a function of the difficulty of seeing the ivory-bill in the dense growth of the forest interior, or perhaps, like many bird species, ivory-bills exploit resources found along the edge such as muscadine grapes and persimmons. While no one would call the ivory-bill an edge species, Ramsey's sightings suggest that it may use resources found along forest edges.

Ramsey is the type of person who will play a prominent role in the future of the ivory-bill. His primary interest is deer hunting, but, as he pointed out, "If we maintain a healthy forest and take care of the land, we also provide habitat for many other species, many bird species." In south Louisiana, the ivory-bill will not expand its numbers without the help of landowners and/or hunting club members like Ramsey. I am hopeful and confident that federal and state authorities will recognize the value in working with these individuals.

In follow-up e-mail exchanges in 2007, Ramsey said that he believes the ivory-bill is still using his land after a brief interruption caused by Hurricane Katrina. He reported that he saw and heard an ivory-bill "a few times" in the late winter/spring of 2007. While the official search in this area was discontinued after Katrina, I believe this is the area from which a photograph of the ivory-bill will be produced.

Jay Boe: Fielding Lewis II?

Jay Boe—an avid duck hunter and outdoorsman I met in April 2005 in Baton Rouge—is the twenty-first-century version of Fielding Lewis. With the amount of time he spends in the woods and swamps of Louisiana, he will one day no doubt chronicle his adventures in his own version of Lewis's *Tales of a Louisiana Duck Hunter.* He has even read Lewis's book and tracked him down to talk about ivory-bill sightings. Lewis's advice to Boe was, "Keep your mouth shut because people will attack you like they did me and Lowery." However, while Lewis initially had deep suspicions about the motives of the conservationists and birders, Boe feels sure that hunters, birders, and conservation authorities can and should benefit each other. In fact, he believes the recent ivory-bill sightings in Arkansas can be a boon to duck hunters. "The Corps of Engineers never helped duck hunters," he told me, "but now that they found the woodpecker in Arkansas, the corps can't touch that area or others where ivory-bills might still be. Duck hunting is totally compatible with ivory-bill conservation. That's why Ducks Unlimited jumped on this issue."

Boe, who spends most of his weekends and vacation time during hunting season in the Atchafalaya Basin, Pearl River area, or some other wilderness area around Baton Rouge, told me he has seen the ivory-bill on five occasions. While that might sound outlandish to some, his knowledge of the woods and the time spent in them, coupled with the fact that Van Remsen directed me to Boe, make me believe that his sightings of the ivory-bill are credible. Like many others who claim to have seen the bird, Boe stated that he has long known what a pileated woodpecker looks like, and that the big woodpeckers he has seen are not pileateds. Also like others, he has furnished clear details about certain features of the birds he saw, such as the yellowish eye, large size, flight pattern, and coloration of the female (all-black crest). Also, because Boe has seen the bird while hunting, when he was being extremely quiet, he was able to see the bird on one occasion at very close range (within five feet in one incident).

Boe is fascinated with the ivory-bill for various reasons, including the bird's majestic appearance, its rarity, its symbolic value, and, now, the fact that so many people—people who have never entered its terrain—claim that it is extinct or nearly extinct. This is perhaps Boe's biggest complaint with the birding community—they don't venture into the heart of the swamp. He says he has spoken with many birders who fear snakes, wild pigs, and men with guns. But, if they really want a chance to see the ivory-bill, Boe said, the birders may have to rub elbows with some of the other creatures that occupy the habitat, such as snakes, bugs, and hunters. Boe related the story of an out-of-town birder who had heard about his experiences and contacted him for information about the birds in the Pearl River area. The birder peppered Boe with safety questions pertaining to wild pigs and alligators. Boe, growing impatient, finally told him, "The most dangerous part of your trip will be the nights you spend in New Orleans!"

When asked why more hunters and fishermen haven't seen or reported the bird, Boe replied, "Actually, there are many informal reports of hunters and fishermen seeing the bird, especially in the Atchafalaya, but like the Fielding Lewis story, no one believes them or they are afraid of what the federal government might do if they have seen it on private land." On the other hand, according to Boe, "many commercial fishermen who you think would see it are in loud boats and take the same routes through the basin day in and day out. Just like game animals such as turkeys, the ivory-bill has learned to avoid those areas." This observation is reminiscent of one by Van Remsen: ivory-bills, although they might not be true game birds, may have evolved, like game birds, to avoid people.

The first time Boe claims to have seen an ivory-bill was in 1982 in the Pearl River Wildlife Management Area. He was squirrel hunting that day and heard a scratching or rustling on the other side of a large tree. Squirrels have a tendency to move opposite a person around a tree, so it made sense for Boe to move as quietly as he could around the tree to get a shot at the squirrel. When he appeared on the other side of the tree, a large, black-crested woodpecker exploded into flight. Needless to say, Boe was startled when the woodpecker flew away. He was only five feet away from the bird and was able to see the black crest and the yellow eyes as it flew away.

Boe's other sightings occurred in 1985 in St. Charles Parish, while he was in the Paradis Canal; in 1988 and 1989, while he was hunting in St. Helena Parish on a large timber tract; and in 2002, while he was hunting near the grounds of the Angola State Prison. Interestingly, Boe claims that after he told the land manager of the St. Helena timberland, the property was logged just six months later. One can't help but wonder if that was a coincidence or part of a pattern that might develop as more ivory-bills are reported in the coming months and years?

Boe was not surprised that the Pearl River search team did not find any birds because they were not sufficiently stealthy. According to Boe, "when I saw a picture of one of them splashing through the water, I knew they would not see the ivory-bill." Boe told me that he had asked Van Remsen how many turkeys the group had seen. When the answer was "not many," Boe knew that the team had put too much faith in past studies that stated the ivory-bill was not shy and was easy to approach.

Boe sightings obviously have convinced him of the ivory-bill's survival, especially in Louisiana. In fact, he thinks that as the momentum brought about by the Arkansas sighting leads to more and similar searches in Louisiana, the ivory-bill will be found in the very near future. When asked where he would search if he had limited time and money, he replied, "the Atchafalaya Basin in general, but the far north around Indian Bayou (north of I-10) and the far south around Franklin in the Bayou Teche National Wildlife Refuge." Other areas he mentioned included the northern Pearl River Wildlife Management Area, including the Bogue Chitto National Wildlife Refuge, and along the Mississippi River north of Baton Rouge around Cat Island National Wildlife Refuge.

While optimistic that the ivory-bill will be found, he does not think the State of Louisiana is doing everything it can to find the ivory-bill. According to Boe, the Louisiana Department of Wildlife and Fisheries budget is financially strapped and therefore unable to fund another expensive, large-scale search like the one that took place in the Pearl River Wildlife Management Area in 2002. Consequently, he suspects that many good reports are being ignored or swept under the rug. While state officials would likely deny this claim, one wonders, given the limited budget of the perennially broke state government (especially after Hurricanes Katrina and Rita), if it may not contain some truth.

Not only is Boe confident that the ivory-bill will be found in the very near future in Louisiana, but he also feels certain that the person who gets a photograph will be a nonbirder, someone like a hunter or crawfisherman who spends a lot of time in the swamps, marshes, and forests. For example, he had a very curious experience with a crawfisherman in May 2005. The fisherman was out on his boat with a very expensive camera. When Boe observed that the "camera was worth more than his house," the fisherman replied that someone with a lot of money had given him the camera in order to take a picture of the big woodpecker, "not the jackass woodpecker [as the pileated is sometimes called in Louisiana], but the one with the big white beak." If outsiders are supplying crawfishermen with cameras, perhaps they are finally starting to take local reports seriously and to realize that outdoorspeople are the ones most likely to encounter the bird in this formidable environment.

The Lord God Bird in Tangipahoa Parish?

I sought out local stories about the ivory-bill through word of mouth, personal referrals from birders and others, and even requests for information in the Baton Rouge *Advocate.* Betty Smith (pseudonym), one of my informants, found me through Smiley Anders's column in the *Advocate.* Smiley's community news/gossip column is read religiously all over south Louisiana, especially by older residents. By making a public request for ivory-bill stories, I was able to reach people outside the academic, professional, and serious birding networks who are usually ignored by the birding establishment. These people can often provide invaluable local cultural and environmental stories. However, Betty's first experience with revealing her sightings to a birding professional was so negative that friends had to talk her into contacting me.

Betty stands out as a perfect example of the often contentious relationship that has developed over ivory-bill reports between the nonbirding public and hard-core birders and ornithologists. Our initial meeting was in Hammond, Louisiana, in May 2003. She is a very kind woman who felt belittled by the ornithologist to whom she reported her sightings of both a male and female ivory-bill within a month of each other in January 2003. Any cautious ornithologist might be likely to greet a report from a perfect stranger with skepticism, but Betty described an exchange with a professor that made her feel as if she had had a dagger plunged into her heart. She had already been intimidated about calling a college professor, and when the expert refused to consider visiting her land to investigate the claims, she felt as though he were calling her a liar.

Betty said the professor on the other end of the line peppered her with accusatory questions. According to Betty, he was not interested in the characteristics she clearly saw, but focused more on what she could not identify. She is not a birder, though she can identify some of the common species on her land—the pileated woodpecker, for example. While she was aware of the buzz about the ivory-bill given the extensive media coverage of the 1999 Kulivan report, she was not out to rediscover the ivory-bill.

After Betty contacted me, we agreed to meet for lunch in Hammond, Louisiana. The forty-minute trip to Hammond is an easy and rather boring drive from Baton Rouge. Still, there was always a bit of anxiety before meeting a total stranger to discuss an ivory-bill sighting. What would she be like? Would she be honest, or could she be seeking some personal benefit by telling me her story? Several state officials involved in the ivory-bill search had warned me that there are a lot of crazy stories and people out there. One story in particular stands out as both hilarious and disturbing. A state conservation official received regular phone calls from an individual who claimed to have seen an ivory-bill. But when the official asked for a photo or some other evidence, the individual sent photos of the mounted specimen from LSU's Museum of Natural Science!

Betty's story begins in early January 2003. While clearing dead limbs from her horse trails in Tangipahoa Parish after a violent storm the night before, she saw what she believes was a male ivory-billed woodpecker. Flying fast and straight from a patch of mature bottomland forest, the bird flew directly over her as she stood in a trail clearing next to a small pond. While her view of the bird lasted only a few seconds, the large woodpecker literally flew only several meters above her head. In fact, the bird flew close enough for her to hear the wind being pushed through its wings. When I stood where Betty had seen the bird, it was easy to understand why the sighting was so brief. The pond and trail form a sort of hole in the forest, with bottomland forest on one side and a pine plantation on the other. Thus, a bird flying over would not be able to see anyone or anything in the hole until it was immediately upon it or them. Similarly, anything that flew over the cleared space would be visible only briefly because of the tall trees that surround in all directions.

There were two features of the bird that stand out in Betty's mind—

the bill and the eye color. According to Betty, the bill was very large and "pearly white." The bill seemed to be "pounded" into the woodpecker's face, with skin "puckers" found where the bill met the feathers. The bill was not the sleek bill of a smaller woodpecker. Betty also provided a vivid description of the bird's eyes. They were yellow, almost glowing. The eye color mesmerized her, and when she retold her story, she became transfixed again. Both the bill and eye color of the ivory-bill differ distinctly from those of the pileated. In fact, Tanner states that the bill is one of the key identification features of the ivory-bill because it varies so much from that of the pileated. The eyes are also important. In photos of live ivory-bills, the eyes appear almost ghostly looking. While this appearance is most obvious in actual photos—especially those taken by Tanner—it is even evident in paintings such as those by Audubon and Wilson.

Betty also described the scene as one of surprise and even shock on both her part and that of the bird. She said that as she stood in the small opening in the forest, the bird seemed to look at her as it flew over the opening, just above her head, as if it were wondering what she was doing in its forest. Several other pieces of Betty's story are worth recounting as well. Betty also claims to have seen a female ivory-bill in her yard about a month after seeing the male. The black-and-white bird (females do not have red crests) briefly perched on the trunk of a large tree in the front of her house. Betty claims this bird was the size of a crow and had white on its back. The only other woodpecker with such markings is the redheaded woodpecker, and Betty claims the bird she saw was much larger than this smaller bird. I at first suspected she had seen a redheaded woodpecker, and its white back had made her think it was an ivory-bill. However, while she could not definitively state whether or not the bird had a black crown, as all ivory-bill females have, Betty insists this bird was large, about the size of a crow. Also, just prior to these sightings, Betty had heard a large woodpecker drumming on one of the dead trees behind her home. While this event is not particularly noteworthy, the pattern and loudness of the drumming are significant. She is adamant that this drumming or rapping was much louder than that produced by the pileated. She also claims that the drumming did not have the rapid-fire pattern produced by the pileated, but instead consisted of short bursts of several blows. She did not see the bird making this noise.

Betty's stories seemed credible enough to warrant a visit to her home. She was a little reluctant because she manages rather than owns the land on which she made these sightings. Several local people, including her boyfriend, discouraged her from bringing outsiders to find the ivory-bill because they feared the federal and state government might take over the land in the name of saving the bird. In fact, according to Betty, her boyfriend said he would shoot the bird if its presence meant property ownership might be in jeopardy. While Betty laughs at these threats, she is not quite sure that they are made in jest.

Professor Vernon Wright accompanied me to Betty's house about four weeks after our initial lunch meeting. Betty was happy to host Wright because I told her he was not nearly as critical as many others regarding ivory-bill sightings. She said she did not want to experience another critical rejection by an academic, but if Wright were more open-minded, he would be welcome to come along. In my evaluation, the property Betty manages has mixed potential as an ivory-bill habitat. There were some nice old oaks and pines, but these were patchy, with young pine and second-growth hardwoods mixed in throughout. Although the property could certainly house any other type of woodpecker, I was a little skeptical as to whether or not there were enough big dead or dying trees to house a pair of ivory-bills. However, after we ventured more deeply into the property on all-terrain vehicles, we did come across larger stands of bottomland forest, especially adjacent to several creeks and sloughs. We visited in the heat of the afternoon, not the most ideal time for seeing wildlife. And, in fact, we didn't see or hear much of anything. However, the forest was obviously home to many woodpeckers. We found numerous large nest holes, as well as boring holes on many of the dead trees. Some of the nest holes did appear to be large enough for ivory-bills, although most were obviously pileated cavities because this species was quite common on the property.

Even though the habitat improved as we traveled around, I was still a bit disappointed by the patchiness of the woods and by the close proximity to pine plantations. The mosquitoes were also ferocious, which probably reduced my open-mindedness and curiosity. Even with bug spray, they swarmed over any uncovered skin. Toward the tail end of our outing, however, we came across a nice patch of bottomland forest next to what appeared to be an old cut-off bayou that had several large nest holes. What stood out most was a single tupelo gum tree being worked over by one or more large woodpeckers. Not only were there many large boring holes, but the bark had been extensively scaled. According to Tanner, scaling is one tell-tale sign of the presence of ivory-bills. In fact, Tanner states that he could walk into a forest and, based on features such as the presence of scaled trees, tell whether or not ivory-bills were present.

When a tree is scaled, the bark of the dead or dying tree has been chiseled or peeled off by a large woodpecker in order to get at the beetles and grubs underneath. Audubon's famous ivory-bill print shows the bird scaling the bark of a tree. Scaling is probably the ivory-bill's most common feeding technique. Therefore, the presence of scaling is a possible indicator that ivory-bills are present. Pileated woodpeckers also scale trees, but they do it with much less frequency than do ivory-bills. Usually the sections of bark that are removed by ivory-bills (and, to a lesser extent, by pileateds) are small pieces about the size of an adult human's hand; however, there have been reports of ivory-bills scaling huge pieces of bark in a single excavation.

The tree in question on Betty's property had been scaled over a significant period of time. Bark had been removed some time ago, but closer to ground level the scaling had taken place much more recently. It appeared as if the birds were moving down the tree. Individuals often claim a tree has been scaled when actually the bark of a dead tree has simply fallen off naturally as the tree continued to rot. The scaling on this particular gum was obviously not natural. In fact, this was the most extensive example of woodpecker scaling I had seen to date. Unfortunately, we found neither feathers at the base of the tree nor any other clues as to what type of woodpecker had scaled the bark.

Betty had the "I told you so" look when she showed us the tree. Perhaps she sensed I was a getting impatient or that we thought the habitat was marginal. Either way, she was happy that we were impressed with the small patch of forest and especially the scaling. In hindsight, and after visiting the forests in the Patterson area where I know ivory-bills exist, the forest on Betty's property seems to have more potential than I originally thought. At the time I had believed, like many others, that the ivory-bill needs huge expanses of virgin timber. Given the recent sightings in Arkansas and the developing story in Louisiana, however, it now appears that ivory-bills can survive in less than perfect habitat. I am not 100 percent certain that Betty has ivory-bills visiting her property, but given some of the signs and her detailed description of the bird, I am not ready to totally discount her claims either.

Pearl River Wildlife Management Area 1999

David Kulivan's sighting of a pair of ivory-bills in Louisiana's Pearl River Wildlife Management Area on April 1, 1999, is the report that really reignited the birding and ornithology communities' interest in the ivory-bill. His sighting also received more attention from professional ornithologists, state and federal conservation groups, and the media than any other past reports of the ivory-bill. Many major media outlets that picked up the story sent reporters to Louisiana to find out more. The story received coverage in major newspapers like the *New York Times*, the *Atlanta Journal-Constitution*, and the *Wall Street Journal*, as well as in the National Public Radio series *Radio Expeditions*.

Kulivan's sighting has also been one of the most, if not the most, controversial. As has occurred in response to other sightings, camps of believers and nonbelievers have formed. Insults have been hurled, and, among some, hard feelings have resulted. I am not certain what prompted so much media coverage of this particular report, when there have been many credible reports over the years. Perhaps the attention was triggered by the detailed description of the birds that Kulivan provided, or maybe it was the fact that Kulivan is a credible source who, as an LSU student, had links to individuals within the birding community like Van Remsen. Whatever the reason, this report resulted in a full-scale search in the Pearl River area in 2002. Sponsored by a major corporation, Zeiss Optics, the search included the placement of high-tech listening equipment to record the bird's call (LSU Museum of Natural Science).

The fact that this report came out of the Pearl River area did not especially surprise many in the birding or ornithology community. Reports of ivory-bills from the area had filtered in for years. And among several elderly people I talked with who live near the southern Pearl River, there was a similar lack of surprise at Kulivan's report. While much of the area has been logged at one time or another during this past century, the Pearl River area remains an expansive, albeit second-growth, wilderness area. Bounded on the north by the Bogue Chitto National Wildlife Reserve, by Honey Island Swamp on the south, and by large expanses of forest in Mississippi to the east, including NASA's Stennis Space Center, the entire area makes up a formidable, often impenetrable, bottomland and swamp wilderness. And yet at the same time, two interstate highways (59 and 10) cut through the Pearl River Wildlife Management Area, and New Orleans is only a short drive away.

Like all the individuals whose reports are included in this chapter, Kulivan was not a serious birder, nor was he out looking for the ivory-bill. He was turkey hunting on a cold spring morning when the ivory-bills entered his life. He resembled a young Fielding Lewis more than a Roger Tory Peterson. The fact that Kulivan is not a birder intensified the criticism leveled at him from the birding community. The fact remains, however, that Kulivan sighted the birds from a vantage point that few birders have experienced—sitting in a wet forest on a cold morning completely dressed in camouflage clothing.

While Kulivan sat perfectly still, a pair of ivory-bills landed on a water oak (*Quercus nigra*) a short distance away. What makes this report even more intriguing is that Kulivan saw a female bird with its distinctive black crest. Skeptics who believe that the ivory-bills that eyewitnesses report seeing are actually pileated woodpeckers often mention the lack of female ivory-bill sightings. This is a legitimate concern. If only males remain, how does the population sustain itself? But Kulivan's report provided a clear and detailed description of a female ivory-bill, including such details as the crest being turned slightly frontward.

Kulivan was a senior at LSU when he saw the birds. On the Monday following his sighting, he shared his story with Professor Vernon Wright. Kulivan was enrolled in Wright's wildlife-management class and said he felt that Wright was the appropriate authority to tell. Upon hearing his story, Wright asked Kulivan to write down exactly what he had witnessed. Wright provided me with a copy of the report in 2003. Below is his report, exactly as Kulivan wrote it:

> On Thursday, April 1, 1999, I observed a pair of Ivory-Billed Woodpeckers (*Campephilus principalis*) on the extreme northern edge of the Pearl River Wildlife Management Area. The pair, a male and a female, were spotted in a water oak at 6:15 a.m. I was struck by

their large size and white patches on their wings, which were clearly evident when they were at rest. They flew from the tree and traveled at tree-top level over an opening in the canopy directly above me. At this point I could see their unmistakable black and white markings below their wings. They landed on a water oak with an approximate dbh of 22 inches. They were now about 10 yards from me. I estimated their length at 20 inches. The female stood at the point where a large branch joined the tree while the male stood on the vertical bole and began to "jump" around and peck the tree; it was very loud. It was also at this point when he (as best I could tell it was the male) began to call. I can best describe the call as a loud, nasal kent. At such a close distance I was able to get an even better look at their physical characteristics. They had very white bills and large crowns—larger and more pointed than the Pileated Woodpecker. The male's crown was red while the female's was black. After several minutes in this tree, they flew behind me and out of sight. However, I continued to hear their call for the next 10–15 minutes.
David Kulivan [signature]
April 7, 1999

Given the detail that Kulivan provided, either he made up the entire story or he saw a pair of ivory-bills. The description and behavior of the birds he saw do not resemble those of a pileated woodpecker at all. And we must first ask why Kulivan would invent such a story? It seems unlikely that he would somehow gain from it personally, financially, or professionally. Kulivan was a serious student, a young man unlikely to toy with his professors—some of whom, such as Vernon Wright, were keenly interested in the ivory-bill. As a hunter, Kulivan would seem less likely to report such a sighting; he might be concerned, like many outdoorspeople, that the report could affect access to a location he frequented. Kulivan has not tried to profit from his story in any way; instead, he sought employment with the organization with which he had interned in college, the National Rifle Association.

While the LSU birding community initially tried to keep news of the sighting quiet, the Internet and birding list serves got out the report of Kulivan's sighting, which eventually spread not only through the Louisiana birding community but also among serious birders nationally. At first the report was met with much skepticism, but as detailed description of the birds emerged, this report struck some as more credible than previous ones, especially since Vernon Wright at LSU served as a character witness for Kulivan. And Van Remsen also lent his credibility to the sighting by reporting that Kulivan provided details not described in guidebooks such as the almost curled crest of the female woodpecker. In the Baton Rouge *Advocate*, Remsen was quoted as saying, "If he [Kulivan] concocted this description, he really had to work at it," and "I am 100 percent sure he is not a fraud." And when Remsen speaks, the birding and conservation communities listen, especially in Louisiana.

As public, media, and scientific interest grew, planning for the first organized large-scale search for the ivory-bill began. The search was led by Remsen and set for January and February 2002. Many individuals had searched the Pearl River area between the 1999 sighting and the organized search in early 2002, but the earlier efforts were not as long-term or on as large a scale as the 2002 search. The search party consisted of six world-renowned birders with a specific interest in woodpeckers. This group spent thirty days in the Pearl River area and the Bogue Chitto National Wildlife Reserve, located to the north. The expedition took place in January and February, the two months when snakes, mosquitoes, and alligators are not as prevalent and the lack of foliage offers better visibility. Also, according to Tanner's study, ivory-bills breed between December and February. Consequently, searchers would be more likely to hear the bird's calls and bill rapping. However, even when mosquito numbers are down, the Pearl River region remains a difficult landscape to slog through. High water, rain, and mud make searching for anything difficult, let alone a pair of fast-flying birds.

The search team posted its daily reports on an official ivory-bill search Web page. Unfortunately, the search did not turn up any hard proof—in the form of photos or definitive sound recordings—of the existence of ivory-bills in the Pearl River region. The team did find some tantalizing pieces of evidence—bark scaling in areas with large trees, large nest holes, and some distant rapping that did not sound like the rapid-fire tapping of the pileated woodpecker. However, this type of evidence is not enough to scientifically prove that the ivory-bill continues to live in the Pearl River region. Some members of the search team remained optimistic after the formal search was over, having been impressed with the large nest holes and bark scaling; others felt there were simply not enough big old trees to support ivory-bills. If the birds were there, these members thought, they were likely just passing through.

Overall, the search team determined that the area and original report was promising enough to justify Cornell University stationing twelve sophisticated recording devices at various locations in the Pearl River region for two to three months in order to capture the distinctive ivory-bill call and double-rapping. Like the search itself, the acoustic devices provided little promising news. Initially, researchers thought the double-rappings had been recorded, but after further analyses, they determined that the sounds were from gunshots reverberating in the swamp.

So, where are we left regarding the ivory-bill's presence in the Pearl River area? David Kulivan has never wavered from his original story. He had seen pileated woodpeckers many times previously, and he still contends that what he saw on that April 1 morning was not a pileated woodpecker. After the search team failed to find any evidence of birds, the momentum to search for the ivory-bill waned until the more recent rumors and the subsequent search in Arkansas. Had they seen a bird or a pair or definitively heard the nasal "kent" call, the search probably would have been lengthened and intensified. That no definitive evidence of the ivory-bill's presence was found, however, made many birders even more skeptical. Many of those who thought the initial report was true have since expressed doubts about Pearl. The apparent shyness of the bird(s) in Arkansas, however, has led others to rethink the skepticism they felt after the Pearl River search. Van Remsen, for example, now believes that Kulivan likely saw ivory-bills, but that the birds either had moved out of Pearl by the time the 2002 search took place or simply avoided areas where they came into contact with people. Vernon Wright remains convinced that Kulivan saw a pair of ivory-bills that spring morning. "They may not live in the Pearl area, but they were using it or were passing through," Wright said. He pointed out that there is a lot of forest around the Pearl in which a few birds could move about, depending on the season and food availability.

While critics may point to this search as final proof that the ivory-bill no longer exists in the Pearl area, we must also keep in mind that the search and listening devices covered a relatively small section of it overall. This was certainly the most thorough search for the ivory-bill up to that time, and its results and the efforts of the search team deserve credit. But does the fact that twelve listening devices failed to record the ivory-bill's call or drumming prove the bird doesn't exist in the Pearl area? Thanks to Nancy Higginbotham, who both helped organize the search and participated in it, I was able to view the locations where the recorders were placed. And while they covered some of the better habitat areas, the geographical area covered was pretty limited. Also, it is important to remember that Kulivan's sighting was not the first solid report from this region. Other highly qualified naturalists such as Nancy Higginbotham are positive that they have seen ivory-bills in this area. Given the vast amount of territory included in the Pearl River region, we should probably not rush to write the ivory-bill's obituary there. And now that birds have been found in Arkansas, Remsen looks back at the Pearl River search and shakes his head about how many things they probably should have done differently such as using more recording devices, spending more hours in the field, and using more discretion while searching for the bird.

Wanted!: $10,000 Reward

Not all meetings I have had with local people regarding the ivory-bill have gone smoothly or provided as many solid details as I would have liked. Some have bordered on the bizarre and even dangerous. A case in point was my meeting and field outing with an individual offering a $10,000 reward for a photo or recording of the ivory-bill.

While most of the individuals discussed in this book are deeply fascinated by the ivory-bill, few have invested more time or money than a man I'll call Mr. Renault of New Iberia, Louisiana. I became aware of Renault's obsession with the ivory-bill through a flier he posted around towns in the southern Atchafalaya Basin offering his reward. Besides posting the offer of reward money, Renault has spent a great deal of his own money on video and recording equipment, boats, and miscellaneous gear needed for exploring swampland. He also generally devotes several days a week to searching for the bird in the Atchafalaya Basin.

The origins of Renault's obsession are unclear. He is not from the Louisiana or the South, having spent much of his life in the Midwest. Even after talking with him at length about the ivory-bill on the telephone in the summer of 2003, I could not discern his motivations, save for a desire to find something extremely rare. Renault is retired from the U.S. Marines and seemed to have the time and resources to look for the bird. He considers himself a serious birder with a great deal of personal interest in other extremely rare birds such as the California condor.

In August 2003, as I had done for meetings with so many other individuals interviewed for this book, I headed west over the Atchafalaya Basin on Interstate 10 to reach New Iberia. As on past trips, as I drove over the basin I was amazed and intimidated by the great green swamp on both sides of the interstate. I had left for this meeting before dawn, and the dark made the vast swamp look even more mysterious and forbidding. I was amazed that such a huge and impenetrable expanse of wildness exists so close to my home. "How can anyone find a couple of birds in that jungle?" I often ask myself as I drive west. And I wonder at the same time whether I want to be in there with the snakes and alligators. As I pass the Whiskey Bay exit, my trepidation focuses more on humans than on wildlife. This is a rather infamous stop, where the bodies of murder victims, including one of the unfortunate victims of the recent Baton Rouge serial killer, have been dumped. And the fact that I was heading into the swamp with a former Marine offering a huge reward for a glimpse of a woodpecker made this trip somewhat unnerving from the start. Who knows how crazy this might get? Renault turned out to be exactly what I expected—buzz cut, camouflage outfit, and machete. A true Marine birder!

Renault insisted we enter the swamp at dawn, just when the birds and wildlife begin to stir. While this makes perfect sense from a birding standpoint, a little voice inside me asked, "You're not really going in there before the sun is up, are you?" We did indeed enter the swamp in the dark, using Renault's ten-foot aluminum john boat, with no running lights. The little voice chimed in again, "Does this guy know what he's doing?" My trepidation was further fueled by the fact that Renault has lived in the area for only a couple of years. As we slowly motored along the edge of a cypress island, the water rushed in over the bow of the boat, and we ran aground on the shallow muddy bottom. The voice was becoming more insistent: "Does this guy knows what he's doing?" The Atchafalaya Basin is a formidable place, a huge, monotonous cypress swamp. Military training or no military training, it is easy to get lost there.

After we disembarked from Renault's small boat to delve deeper into the swamp via a canal, it became pretty clear this was going to be a physically challenging day. We immediately set off on foot, heading west through the cypress–tupelo gum swamp. It wasn't the walking that was so challenging; it was slogging through knee-deep water and mud. Renault politely asked if I was OK while he also huffed and puffed through the tangled swamp. If the ivory-bill has remained hidden from the birding public, this is the type of landscape that could have helped to conceal it. While there is certainly no shortage of boaters in the area, very few people wander over such terrain on foot.

The forest, though rather young, appeared to contain appropriate tree species. It was a nice mix of cypress and tupelo gums. It was unclear exactly when the area was last logged. There were few stumps, and those that remained were badly decomposed. Yet the trees were not huge, meaning that the heat, humidity, and standing water must have quickly rotted the dead stumps. I estimated the forest was around sixty years old, with some of the hollow giant cypress trees, which were spared by the loggers, being much older. There were few signs of people. A few very old markers left behind by landowners—a rusty crawfish trap and a few beer cans—were all we came across to indicate any human presence. The lack of trash is one tell-tale sign that few humans visit this place. Sadly, many waterways and forests I have visited while writing this book have been strewn with trash. It seems that while Louisianans love their outdoor activities, many struggle with the idea of carrying their trash out. One indirect sign of a human presence was the large number of feral pigs in the swamp. These pigs were released in many areas of the South decades ago to serve as game for hunters. While they are occasionally hunted, their population is not under control, resulting in a great deal of pig-induced environmental damage in some southern landscapes. According to Renault, the pigs do eat water moccasins. While I was not sure if this were true, the moccasins we were seeing along the way made me wish there were more pigs in the area.

During the walk, Renault delivered some stunning news. Since my first telephone conversation with him a month earlier, during which he confirmed that he is offering a $10,000 reward for a photo of an ivory-bill, he claimed to have actually seen a female bird with three fledglings. He claimed that he spotted these birds in the same forest we were visiting together, but on higher ground that contained a larger proportion of sweet gum and oaks rather than the lower-lying cypress-tupelo forest. As he described his sighting, he almost became overwhelmed with emotion. While I don't think it is my place to challenge his sighting, the forest we were walking in seemed to offer few signs that it could support big woodpeckers. But since Renault insisted that he saw the birds on higher and drier ground, we headed west in search of this "upland" forest. It was very, very slow going. Knee-deep water, knee-deep mud, waist-high weeds, hordes of mosquitoes, and hidden cypress knees made walking an exhausting adventure. It's hard to look for a moving needle in a haystack when you have to constantly fight to maintain your balance. Also, the forest was full of water moccasins. While I generally find snakes fascinating, I try to avoid water moccasins. Unlike most snakes, they don't flee when approached, but boldly hold their ground. You either walk around the moccasin, or you get struck. By the end of the day, we had seen six water moccasins and one copperhead, a less aggressive but still poisonous swamp-dwelling snake. Interestingly, the moccasins were never in the water. Instead, they were always sitting on a dry surface such as fallen tree above the water. These locations made them easier to spot but also made walking even more dangerous because they often rested in areas where I wanted to step. As we continued to walk, I continued to have my doubts about this habitat—even if Renault had really seen ivory-bills on the higher ground, surely they would be using the very forest we were walking in. And there were no signs of ivory-billed woodpeckers.

We walked from 6 a.m. until about noon, due west, according to Renault's compass. We never did find any significant upland forest, or at least none that could have possibly supported a family of ivory-bills. All we found was a bit of dry land with a few scrubby oaks and gums. Needless to say, I became a bit skeptical about his reported sighting. Maybe, just maybe, a lone bird could live in the area, but not a family with three fledglings. And not only was the habitat not as he described it, but about an hour into our trek, Renault compounded my doubts about his identifications by announcing he was color-blind. This was one more piece of a story that wasn't adding up.

After many hours of walking without finding any suitable habitat, and with our water running out, I strongly suggested we start heading back east to the boat. Renault had told me to plan on a three- to four-hour hike in and out, so my food and water supplies were minimal. I had intended to examine the habitat, not to search for the bird. It soon became apparent to me that he had no idea where we were. Nothing looked familiar, yet everything looked the same. He assumed if we walked due east we would run into the bayou where we left our boat. From there it would be just a matter of walking a short distance north or south until we came across the boat. By midafternoon, the swamp had become very hot and humid. My rubber waders were turning my feet into hamburger, and I was becoming not only tired and thirsty but also angry. Flooded cypress swamps in the middle of the Atchafalaya Basin are no place to be lost. I had assumed that he had visited the area many times and would have a better understanding of the lay of the land. After we walked for a while, Renault told me he had "gotten lost in this area a lot, but never like this." It became crystal-clear that this "local story" was not really that of a local, and it certainly did not reflect any sort of in-depth local knowledge of the landscape.

Several hours later it was getting late in the day, and we were hopelessly lost. I actually began to lose interest in the idea of finding the ivory-bill. We were both exhausted, and trying to stay alert for more water moccasins, I kept my eyes more focused on the water at my feet than on the forest canopy. By this time, Renault had lost all credibility in my mind by misidentifying common trees and birds and having absolutely no idea where we were or where he saw the female bird and three fledglings. Under these conditions, I actually felt myself beginning to side with the experts who dismiss the many reports they get from locals regarding ivory-bills. I gave voice to the doubts I had begun to feel: "No wonder few experts take these ivory-bill reports seriously. Getting lost in a swamp in knee-deep mud is what they'll end up with!" At one point in what had begun to seem like a death march, we came upon a levee. Renault immediately stated it was a buried pipeline (which it was not) and that if we followed it in a particular direction we would find our way out of the swamp. I asked repeatedly whether he was positive if we walked in direction *A*, we would reach location *B*? "Yes, yes, yes," he replied. Once again, he was completely wrong. We ended up basically at a dead end on the open waters of the Atchafalaya—we had been heading in exactly the wrong direction!

We eventually found our way out of the swamp with the aid of a signal fire and a friendly local Cajun fisherman, who of course thought we were both crazy for traipsing through the area without a GPS unit. While the swamp we visited was remote and rarely visited by humans, it contained few signs of large woodpeckers. I heard a few pileated calls but saw no large nest holes and no bark scaling on any of the dead trees we came across. The area was a very nice example of a maturing cypress–tupelo gum swamp, and no doubt at some point in the past one hundred years ivory-bills had lived in or visited this forest, but I felt certain that, in 2003, there were no ivory-bills living in it.

Renault probably saw a pileated woodpecker. He had become so dedicated to finding the ivory-bill in a relatively small area that he probably convinced himself he saw the bird(s). In some ways, it is easier to convince oneself the bird exists than to admit defeat. One aspect of Renault's reported sighting really bothered me, though, and made me wonder if he had ever seen anything at all. Why, if he was so dedicated to this search and so prepared, had he forgotten his camera on the day he claimed to have seen the bird? According to Renault, he had been out looking for snakes and hadn't thought to bring it along. I find it hard to believe that someone dedicated enough to finding the ivory-bill to offer a $10,000 reward would forget his camera when he goes into what he deems prime ivory-bill territory.

Chance Encounters, Fleeting Memories

In addition to the above stories, many other individuals have shared information with me about past encounters with ivory-bills. Some, but not all, of these encounters are quite old, so details are rather scarce. However, they are worth retelling because they do in some cases indicate that the ivory-bill hung on in Louisiana longer than was generally acknowledged by the experts of the day such as James Tanner. Most stories relayed to me came from people who had chance encounters with the ivory-bill while they were out hunting or fishing. However, other individuals claimed they regularly saw the ivory-bill in certain areas year after year. Most of these reports emanated from the Atchafalaya Basin proper, the far southern section of the swamp in general around Franklin, and the Pearl River region. However, a couple of these sightings occurred in northern Louisiana, with one particularly intriguing report coming from the Tensas River National Wildlife Refuge.

The Singer Tract story was relayed to me by Newty Jeansonne in a telephone conversation in the fall of 2003. Like Betty, whose story was told above, Jeansonne saw my request for information pertaining to the ivory-bill in Smiley Anders's column. According to Jeansonne, he saw three to four pairs of ivory-bills while deer hunting in the Singer Tract not long after World War II. The World War II date is significant because the Singer Tract was logged in the early 1940s (in part, it was claimed, because timber was needed for the war effort). Jeansonne was not sure of the exact year, estimating it to have been in the late 1940s or early 1950s, but he is sure that it was after World War II. According to Tanner's study, the entire area was largely destroyed. Tanner believed the ivory-bills did not survive the destruction. Yet Jeansonne claims that he saw the ivory-bills in the vicinity of the Singer Tract after the logging took place. He did not, of course, claim he saw the birds in a logged-over area. Instead, he spotted them in a forested area just south of the main logging operations where he was deer hunting. This is important because a few birds may have survived in the area. Not all of the land around the Tensas River was logged. Since there were still forest tracts in the area, perhaps the ivory-bills, when forced out of Singer, moved around the region, forest island to island.

Jeansonne is sure about what he saw. He has hunted a great deal over the years, and like so many folks with whom I have spoken, he knew prior to his ivory-bill sighting what a pileated woodpecker looked like. And, like so many others whose sightings are discussed in this book, he said the birds he saw were not pileated woodpeckers. "I was sitting there in the woods waiting for a deer, when three or four pairs of these huge woodpeckers flew in. They were jumping around on the trees, making a big racket. I had never seen woodpeckers like those before," he said. And he has not seen them since.

Were these birds a family group of ivory-bills, including recently fledged birds? We will never know. Or were they the very last ivory-bills to survive the destruction of the Singer Tract? Perhaps these birds dispersed soon after to other areas such as north Arkansas. We can only speculate. But the sighting is important because if this was indeed a group of ivory-bills, it indicates they may have successfully dispersed from the logging zone into surrounding forested areas such as the Atchafalaya Basin. The area in question is not that far from the Mississippi River bottomlands, an area that in the 1940s and 1950s still contained some decent pieces of bottomland forest (albeit increasingly small and isolated, given logging and agricultural expansion), both on the Louisiana and the adjacent Mississippi side of the river. We can only speculate about where these woodpeckers came from or if they moved on. However, if we accept that they were in the Tensas/Singer area after 1945, it is possible that some ivory-bills survived. While the odds are long, these few pairs might have been the parent birds to some or even all of the birds reported more recently in areas such as the Atchafalaya Basin, Yazoo Delta in Mississippi, and the Pearl River region. We do not know how far ivory-bills fly in search of food, a mate, or new territory. While it seems unlikely, it might not be impossible for birds to disperse over time into distant areas.

Moving chronologically toward the present day, I was told of two separate reports of ivory-bill sightings in the Atchafalaya Basin in the early 1970s. Both reports were from the vicinity of Duck Lake, and the people making the reports did not know one another. Both individuals reported, in telephone conversations in September 2003, seeing a lone male ivory-bill. One sighting occurred in the spring and was reported by someone on a fishing trip; the other took place in the winter and was reported by a person on a hunting trip. I have grouped the stories here because they are so similar that the first thought that came to my mind was that the reports described the same bird. Both of those making the reports were experienced outdoorsmen who claimed they knew that the bird they saw differed from the pileated woodpecker. Neither man had ever seen the bird before, and neither saw it after his single sighting.

I suspect, however, that at least one of these men may have seen the birds more recently than he let on. He was quite wary of talking to me, initially suspecting I was a "Fed." While he opened up a bit after we talked a while, he made it clear that he did not want to draw the same sort of attention to the Atchafalaya Basin that the 2002 Pearl River search had attracted. I assured him that a thirty-five-year-old report would not generate that type of response. I wondered if the reason he acted so secretive was that he had recently seen the woodpecker. Given how long it had been since the reported sightings, both individuals gave surprisingly accurate descriptions. Both mentioned the large white bill and size of the bird, while only one described it flying straight and fast—"like a duck." These are all tell-tale characteristics of the ivory-bill.

I include these older sightings for one simple reason: I am convinced the ivory-bill survived the destruction of the Singer Tract and then dispersed to other areas. Or it may have already been present in small numbers in locales such as the Atchafalaya Basin. While Tanner was convinced the woodpeckers he studied were the last, I wonder if this belief wasn't in part driven by nostalgia for the majestic Singer Tract that had been destroyed. Or perhaps there was even a bit of academic egotism in his claim that he had studied the last ivory-billed woodpeckers. Tanner was surprisingly hostile to individuals who claimed to have seen the ivory-bill. For example, he openly questioned the credibility and honesty of John Dennis, who claimed to have seen the ivory-bill in the Big Thicket in east Texas. It remains unclear why Tanner was so unbending in his belief that the ivory went extinct in the 1940s.

There are many other more recent and just as fleeting encounters with the ivory-bill in Louisiana. For example, in February 2001, a former LSU graduate student was traveling north on Route 83 (southwest of Franklin, Louisiana) when a large woodpecker caught her attention as it landed on a roadside telephone pole. While her sighting was brief, she was convinced the large bird was not a pileated woodpecker. Like so many others, she claimed to be familiar with the pileated from her childhood in a rural, wooded area in Pennsylvania. Vernon Wright showed me the letter she sent to him that contained her description, and it was very accurate. The significance of this sighting is that it occurred in the vicinity of Fielding Lewis's sighting, as well as those of some of his cohorts from the Franklin area. It seems likely that a small population of ivory-bills continues to survive in the swamps just south of Franklin. And if a few pairs of birds have bred, past fledglings probably left in search of new territory. They could easily move across LA 90 and into the Atchafalaya Basin because there are a few small areas where only the highway divides the bottomland swamp forest.

The fact that this individual was already familiar with pileated woodpeckers coupled with the proximity to Fielding Lewis's and others' sightings would seem to support Wright's theory that a small breeding population dispersed into past territory such as the Atchafalaya Basin or Pearl River area. I doubt, as some have suggested, that the ivory-bills reported over the years are just a couple of very old birds that have hung on. If these sightings around Patterson, the Atchafalaya Basin, and Franklin are legitimate, as I believe them to be, then this small group of ivory-bills is almost certainly breeding. It is unlikely they are producing fledglings in large numbers, but thirty years of consistent reports of ivory-bills in the Franklin area would strongly suggest they are breeding.

If there are a few Lord God Birds around the Franklin area, one might reasonably wonder why so few photographs of them have been produced. In addition to the difficulty of photographing a fast-flying bird moving through the dense forest, this area is, as Fielding Lewis told me, the wildest landscape left in Louisiana. Only small numbers of hunters and fishermen enter it. Also, and perhaps more important, most of the area is private land owned by oil and gas corporations; therefore, few people can gain access to the forests and marsh to search for the birds. For example, without help from locals, there are few points to launch a canoe into the swamp. According to Fielding Lewis, in 2002 he guided two birders associated with the Pearl River search to the area where he took his 1971 photographs. As they entered they area, the birders changed their minds because they did not want to get wet, Lewis said. The two individuals told Lewis they would return to the area later when they were better equipped. Several years later, Lewis is still waiting for their phone call.

The next story involves two fishermen, two ivory-bills, and a close call. In about 1986, a father and son were fishing in the Grand River area in the Atchafalaya Basin, moving through the water under a canopy of trees, when they heard a large woodpecker hammering a tree above them. Their story was relayed to me in a conversation with the father in the fall of 2004. They were able to identify a male and female ivory-billed woodpecker. According to the father, who requested anonymity: "The male was higher in the willow tree than the female from looks, and as I remember, the male was in full plume, really a magnificent bird. We were not concealed except by the canopy of trees in the slough. My son and I were fishing—very quietly—I'm a slow fisherman and was using the trolling motor. We heard [a] sound and backed off to get a better look at what was making extra-loud noise. We then saw the birds—and thanked God we backed off because the male knocked off darned near the top of the tree—it fell within 10 feet of the boat—a branch 4 to 5 inches across (it was dead, of course) and 4 to 5 feet long. I think the noise scared the birds more than we did." I know the father of the team quite well. He is an experienced outdoorsman, and I am completely confident that he and his son saw a pair of ivory-billed woodpeckers.

Another interesting story is provided by Dean Wilson of Bayou Sorrel near Grand River, which sits on the eastern edge of the Atchafalaya Basin. Wilson came to Bayou Sorrel sixteen years ago to make his living as a commercial fisherman. He was drawn to this area because "it's one of the last places you can live off the land." His story is interesting not only because he claims to have seen ivory-bills twice but also because, having been born in Spain, he is not a typical resident of Bayou Sorrel. When asked how he fit in to the small community, he said, "It was tough at first. Someone actually fired a gun at me. But now people are used to me, so it's better."

When I left the main roads on my drive to meet with Wilson in July 2004, I felt as if I were entering another world. Commercial fishermen stared as I drove by. The commercial fishermen in the Atchafalaya Basin have struggled greatly during the past couple of decades as farm-raised catfish and crawfish have taken a bigger market share away from the traditionalists. According to Wilson, even the best fishermen today must struggle to survive, or, as he put it, they "have to be very good to be very poor." As a result, in places like Bayou Sorrel, the traditional fishing culture has been decimated, leaving poverty and bitterness in its place. It is hardly where you would expect to find a Spanish-American commercial fisherman turned birder.

Wilson was not a birder when he sighted the ivory-bill. Instead, as with many others who have reported seeing the bird, his first sight of the magnificent ivory-bill piqued his interest not only in that species, but in other birds as well. He first saw what he believes was an ivory-bill in his front yard in 1993. He said that he had seen plenty of pileated woodpeckers while he was fishing in the Atchafalaya. But this bird, according to Wilson, was different. "It was larger, the bill was big and white, and its crest was much sharper than the pileated." It was also beating the tar out of a stump in his front yard. Between drumming on the dead wood, it also scaled off pieces of bark in search of insects. At the time, Wilson knew nothing about ivory-bills scaling bark off dead and dying trees. Since he was not looking for an ivory-bill, he was not transferring ivory-bill characteristics to a pileated woodpecker after having read a magazine article or field guide. But he did know, as he witnessed the bird's powerful drumming and scaling for several minutes, that this was no pileated woodpecker.

Wilson had to wait another nine years, until July 2002, for a second sighting of the ivory-bill. He was in the Atchafalaya Basin roughly west of his home, and this time he believes he saw a pair of birds. As his boat turned around a bend, two large woodpeckers with white on the trailing section of the wings flew away from him. Although he didn't get as clear or long a look at these birds as he had with the stump pecker nine years earlier, he said he is confident not only that these woodpeckers were larger than the pileated, but also that they flew straight. He, like others I spoke with, reiterated: "Those birds were not pileated woodpeckers. I see the pileated all the time, and those birds were different."

These experiences changed Wilson's life. He had always been interested in the environment, but the ivory-bill encounters led him to become even more involved in conservation issues in the Atchafalaya. Today, in addition to fishing, he serves as the Atchafalaya Committee chair for the Sierra Club and as the Atchafalaya basinkeeper. A basinkeeper acts as an environmental advocate for a particular location/environment; the position was established by the Waterkeeper Alliance, an organization whose main interests are water quality and the ecological health of aquatic systems. As basinkeeper, Wilson also provides environmental outreach such as giving talks to schools about the value of the Atchafalaya Basin.

An even more recent fleeting sighting of the ivory-bill in the Atchafalaya Basin was reported by Ron Boustany, a natural resources specialist with the U.S. Department of Agriculture. Boustany was flying over the Atchafalaya Basin on September 7, 2005, in order to survey damage caused by Hurricane Katrina. As he passed low over the Bayou Sorrel area, he saw a large woodpecker flying above the trees with trailing white feathers on its wings. Boustany is a serious biologist who, like others who work in southern Louisiana, knows the wildlife. Even though he saw it briefly from a small plane, with the bird cruising along the treetops, he was certain that the large woodpecker he saw was an ivory-bill. A ground reconnaissance team entered the area after the sighting, and although they did not see an ivory-bill, they identified a number of trees that had been scaled as well as habitat that could support the ivory-bill.

While it is impossible to determine indisputably if these sightings were actually ivory-bills, the northeastern Atchafalaya Basin and the Franklin-Patterson area are significant because of both the number of reports of ivory-bill sightings and the credentials of some of the reporters. Perhaps the northeastern Atchafalaya Basin, like Franklin and other locations including the Pearl River region, should be considered "hot spots" for ivory-bills. And while some may question the validity of these and other reported sightings, the descriptions and consistency of these reports over the years have been impressive. Although we should not take these sightings at face value, the individuals described here—hunters, fishermen, field biologists, and outdoorspeople—are the ones who will likely lead the birders and ornithologists to the ivory-bill.

5

FUTURE SEARCHES

The decline of the ivory-billed woodpecker over the past 150 years cannot be blamed on historical limitations of its range. The ivory-bill's territory once stretched from coastal North Carolina to eastern Texas and from the Gulf of Mexico to the lower Ohio River and up to St. Louis on the Mississippi (map 3). Not all of the environments within this vast space were suitable ivory-bill habitat. As one moved to the northern limits of its territory, the birds would have been confined to the areas in close proximity to the Mississippi, Missouri, and Ohio rivers, where bottomland hardwood forests were dominant. And even in the heart of their territory in the Deep South, the birds likely avoided the vast longleaf pine forests that once dominated the Coastal Plain—although there has been speculation that the ivory-bill exploited dead and dying pines adjacent to bottomland forests in search of food (Jackson 2004). Upland areas in the South such as the Ozark Mountains and the lower Appalachian Mountains were also largely devoid of potential habitat.

Even with these geographic restrictions, the ivory-bill did not lack potential territory. Every major river system in the South, from the largest, such as the Mississippi and the Ohio, to small bayous and oxbow lakes, contained potential habitat. Before the construction of massive flood-control structures such as the levee system along the Mississippi River, large areas of the Deep South would have been seasonally flooded or contained standing water. This flooding created the right conditions for the bottomland forests, which in turn created the right conditions for species such as the ivory-billed woodpecker.

However, massive, contiguous tracts of bottomland forests no longer exist. So even though the ivory-bill once inhabited vast geographical spaces in the South, the destruction of their habitat means that today's searches for the bird can be much more selective. (See Dennis [1988] for a comprehensive inventory of the South's great remaining swamps.)

Also, given the restrictions of time and money for ivory-bill field investigations, searchers should concentrate on a handful of areas.

Now that the ivory-bill has been found, the next question becomes where the scientific community should begin to look for more of the birds. In 1986, the U.S. Fish and Wildlife Service organized a meeting at LSU in Baton Rouge that brought together about twenty ivory-bill experts, including woodpecker dignitaries such as James Tanner, Lester Short, Bob Hamilton, and Jerome Jackson. Other attendees included Tommy Michot and Wylie Barrow, who shared their notes and memories of the meeting with me. The group developed a priority list of places to search. The group's number-one search location was the Atchafalaya Basin, followed, in this order, by the Santee / Congaree River areas; the Tensas River National Wildlife Refuge; the lower Altamaha River region in Georgia; the Tombigbee River area in Mississippi and Alabama; the Yazoo Basin in Mississippi; the Pascagoula River region in Mississippi; the Apalachicola River area in the Florida Panhandle; and the Big Thicket area in east Texas. Interestingly, the White River area in Arkansas, where the ivory-bill was found in 2004, did not make the priority list.

I do not believe the ivory-bill exists in large numbers anywhere. But I am convinced, based on my conversations with the wildlife professionals and the outdoorspeople whose stories I have collected in this book, that the birds still exist in a few places in the South. I am especially convinced there are two and perhaps even three locations in southern Louisiana that house a few ivory-bills. These populations are likely isolated and small, making their detection difficult. Given the critically endangered status of the ivory-bill, searches should take place in locations where birds were known to exist in the early twentieth century (into the 1930s) and where sightings of birds have been reported in the past several decades. These areas often overlap, but not always. Since resources earmarked for field expeditions are usually scarce at the state, federal, and university levels, searches should focus on the areas of overlap.

My priority list of search sites would look somewhat different from the one proposed by the 1986 meeting participants, though Louisiana overall, and the Atchafalaya Basin specifically, would top my list, as it does theirs. However, unlike the 1986 group, I would place the Big Thicket region and the Apalachicola National Forest region higher on the list. The Santee River and Congaree Swamp areas in South Carolina deserve renewed attention because of their habitat.

How the searches are conducted is critical. Certainly teams of experts should assess an area's forests on the ground and via remote sensing and aerial photos. At the same time, interviews with local people should be conducted in promising areas. As the interviews with Louisiana outdoorspeople collected in this volume illustrate, local people have knowledge of, access to, and experience in nearby forests that should be made a critical component of any search and preservation strategy. Less-intensive methods could also be used, such as acoustic recording devices to record any ivory-bill calls without actually spending large numbers of man-hours in a forest.

In the hope of providing information that may lead to the timely rediscovery of the bird in other locations, I outline below several locations that have promise as potential ivory-bill habitat.

Southern Louisiana

The southern edge of the Atchafalaya Basin near Patterson was the location for the formal 2005–6 search because of the recent sightings and audio identifications of the ivory-bill from that area described previously. However, sightings of the ivory-bill in this region aren't new. Forested areas between Avery Island (on the west end) and Morgan City (on the east end) south of LA 90 have produced many outstanding ivory-bill reports over the past few decades, the most famous being the photos sent to George Lowery by Fielding Lewis in 1971. However, there have been several very credible reports from this region since the 1970s. After the recent sightings in the Patterson area in 2005 (pre–Hurricane Katrina), area resident Scott Ramsey reported seeing and hearing ivory-bills on his property in 2007, and birder Jay Huner reported a sighting in June 2006 (shortly after the official search ended). Based on my conversations with locals in this area, there have been few large-scale environmental changes in the area's forests since Lewis's photographs were taken. Mature trees still dominate, although there are more dead and dying cypress trees due to saltwater intrusion and hurricanes. While these dead and dying trees are part of the larger environmental tragedy for the coast of Louisiana, they also provide habitat for ivory-bills and might be one factor responsible for many reports of ivory-bills in this area in recent decades.

This area is not all wilderness. Sugarcane farms dot the landscape, and gas and oil exploration has cut canals through the marsh and forests. But mature forests still provide the majority of ground cover, especially just south of LA 90. This is an intriguing landscape with bottomland forest located on slightly higher ground, cypress swamps in the low-lying areas, and marshland as one moves south toward the coast. Also found in this area are salt dome hardwood forests. These ecosystems are dominated by hardwood species on uplifted land, so that they are essentially forested islands surrounded by wet marshlands (Martinez 1991). E. A. McIlhenny, of Tabasco hot sauce fame, reported ivory-bills in the Avery Island salt dome forest. He stated: "By 1918, black bear and Ivory-bills were unusually plentiful in the forests of Avery Island" (McIlhenny 1941:583). Soon thereafter, though, the Avery Island forest was logged, and the birds and bears disappeared.

Today, Avery Island as well as other salt dome forests house mature forests again. Could these forested islands serve as refuges for ivory-bills? Because of the secrecy surrounding Fielding Lewis's photographs, this area has not seen any organized searches until recently. Home to the 9,000-acre Bayou Teche National Wildlife Refuge (created to protect the Louisiana black bear), the area is, in fact, rather off the beaten birding path in Louisiana. Should an ivory-bill be documented in the area, Bayou Teche National Wildlife Refuge could serve as a core zone for any ivory-bill management plans and protection (U.S. Fish and Wildlife Service, Bayou Teche National Wildlife Profile).

Access to most of this terrain is difficult because much of it is dry. With standing water seasonal and shallow in most bottomland forest, boats usually cannot be used there to get around. At the same time, the forest is mature with dense undergrowth containing palmettos, which, along with the large number of water moccasins and canebrake rattlesnakes, makes it a challenging environment to slog through. However, because Bayou Sale and the Intracoastal Waterway run through some of the area, some searches could be conducted from the water.

The levee adjacent to the Intracoastal Waterway is relatively high and provides an open view of the treetops and forest edges. It could provide an advantageous location at which to station observers. If ivory-bills exist in this area, it is highly likely that they travel throughout the forests south of LA 90 simply because the overall amount of mature forest is not that large, at least compared with the Atchafalaya Basin to the north.

In addition to the southern edge of the Atchafalaya Basin, the area just north of LA 90 should be considered for a more intensive search. Numerous reports of ivory-bills, including Tommy Michot's sighting, have come from the Duck Lake area in the southern section of the Atchafalaya Basin. These could even be the same birds that have been reported just to the south, on the other side of LA 90. Given that forest encroaches on the highway in a few areas, it would not be that difficult for ivory-bills—especially birds that require large tracts of habitat—to travel between these two regions.

Grand River Area

The second area that I would recommend searching is the northeastern Atchafalaya Basin. The area between Bayou Pigeon and Grand River should be the focal location. This is the area where Bob Hamilton heard and saw an ivory-bill in the late 1970s; where two fishermen (whom I know personally) clearly saw a pair of birds in 1986; and where Ron Boustany, a U.S. Department of Agriculture biologist, saw an ivory-bill from a small airplane in September 2005. There have been other very competent birders and ornithologists who caught glimpses of what they believed to be an ivory-bill in this general region. A couple of these sightings occurred on the I-10 bridge through the Atchafalaya, to the north of Grand River. This area, where some logging continues today (see figure 7), is not pristine, but it remains a wild location with some mature timber. Also, it is connected to the much larger Atchafalaya Basin, allowing local birds who may roost or feed in the area access to a much larger resource base. Overall, the Atchafalaya Basin contains roughly 600,000 acres of bottomland and swamp forest, the single largest such environment in the United States (Demas 2001).

The Atchafalaya Basin too, even though it has been altered by various large-scale water-control projects and has been logged, remains a wilderness area (Sierra Club Profile of the Atchafalaya Basin). While some might be tempted to dismiss the basin as a cut-over area where ivory-bills could not possibly survive, as you drive through it on Interstate 10 and look out over the vast forests, it becomes apparent that, even with past and present human manipulation, this is still a wild place. The basin was not clear-cut at the same time in all locations, and enough old timber and imperfect trees (from the loggers' perspective) were left in and around the area to support ivory-bills. Skeptics have questioned why, of the many hunters and fishermen who enter the basin, none has yet reported sighting the bird. My research indicates that some outdoorspeople have reported it, but their reports weren't accepted as credible. And there may be fewer local reports because large areas within the basin are owned by private landowners or hunting clubs who fear the government intrusion that could result from an endangered species being found on their property. In my conversations with local people, I heard repeatedly that no one wanted a "southern spotted owl." These leased or private parcels in the Atchafalaya Basin are essentially off-limits to outsiders who might be looking for a rare bird. The fact that much of the basin is private land poses great challenges for conservationists if the bird is found.

Nevertheless, because many knowledgeable individuals have reported ivory-bill sightings in this area, it merits closer scrutiny. One intriguing aspect of these sightings is that many have come from specific areas within the Atchafalaya Basin, such as the Grand River area in the northeastern section. No fewer than six credible reports of ivory-bills have come from that area during the past thirty years.

Pearl River Area

Another organized search for the ivory-bill should be conducted in the Pearl River area. Although some remain skeptical about David Kulivan's 1999 sighting there, I believe that the bird's rediscovery, coupled with the reports of other past sightings in the area that I collected while researching this book, justify accepting Kulivan's report from the Pearl area as legitimate. The Pearl River region remains a vast, little-explored area that is connected on the north and east to other large bottomland wilderness areas (Louisiana Dept of Wildlife and Fisheries, Pearl River Wildlife Management Profile).

Overall, about 185,000 acres of forestland (in various stages of growth) are found in and around the Pearl River region. This includes the Pearl River Wildlife Management Area; the Bogue Chitto National Wildlife Refuge on the north; NASA's Stennis Space Center on the east; and Honey Island Swamp on the south. Given this great territorial expanse, is it any wonder the 2002 search team did not find the ivory-bill? While the 2002 search by Cornell and LSU did cover a large amount of territory and was carried out by many experienced ornithologists, the vast Pearl River area remains largely unexplored. And, as previously mentioned, Van Remsen at LSU said that, in light of the later Arkansas expedition, the 2002 search team was not secretive enough. A small woodpecker population that moves throughout the Pearl River area as well as the adjacent wilderness areas could easily avoid detection. There simply have been too many solid reports of ivory-bills to doubt that at least a few birds still exist in the Pearl River (or at least use it during certain times of the year).

Like the two previously mentioned areas, the Pearl River region is a difficult landscape in which to search for a handful of birds that are constantly on the move. Certainly audio recording devices could be redeployed in more areas to search for the bird without a constant human presence. Another effort that might lead to a more cost- and time-efficient search is to foster communication with local hunters and fisherman. If an atmosphere of mutual trust and respect could be developed between local outdoorspeople and the birding community, public meetings seeking input from local outdoorspeople might increase the likelihood of their reporting possible sightings. The Arkansas state government now offers a reward for information about the ivory-bill, an incentive directed at local people. In areas used extensively by the nearby residents, a community-based approach makes sense.

Tensas River National Wildlife Refuge (Former Singer Tract)

There is a great deal of irony, unfortunately, in suggesting that a search for ivory-billed woodpeckers should take place in and around the former Singer Tract, in what is now the Tensas River National Wildlife Refuge, because it was largely cut over during the 1940s (U.S. Fish and Wildlife Service, Tensas River National Wildlife Refuge Profile). Therefore, if ivory-bills were to reappear in this area, it would be a testament either to their ability to hang on in degraded environments or to their ability to leave, travel long distances, and then return to a specific location as the forest matures, or both.

A few interesting reports of ivory-bills have come from in and around Tensas during the postlogging phase (late 1940s), and the now protected forests have regrown into an impressive secondary forest. For example, in 1981, two respected birders reported hearing an ivory-bill call while visiting the area, and local people occasionally have also claimed throughout the postlogging phase to have seen ivory-bills (Heinrich and Welch 1983). Reports of ivory-bills have also come from nearby areas in Mississippi. In addition, while the Tensas River National Wildlife Refuge, like so many large forest patches in the lower Mississippi Valley, is largely an island of trees among cropland, the reserve isn't that far from other patches of forest, making it possible for ivory-bills to reenter the area if they did leave after logging. Of course, this assumes the ivory-bill would cross large open areas to reach new territory. While past studies have documented that the ivory-bill will travel long distances in search of food or nesting trees, it is not well understood if ivory-bills would move through open territory like cotton fields, or if they need forest corridors. If the birds are found in Tensas River National Wildlife Refuge, it will be established that they will cross open fields, or at least use degraded forest corridors.

The Tensas River National Wildlife Refuge has regained some of its biological and botanical luster that was destroyed in the 1940s. If a current visitor knew nothing about the past logging, he or she would probably be surprised that the area was essentially leveled just sixty or so years ago. Today, the 70,000-acre Tensas River National Wildlife Refuge contains some very nice tracts of maturing second-growth forest. Of this refuge, 54,000 acres are covered in forest. Given its size, the refuge could probably house several pairs of ivory-bills again. The forest will likely never regain its former structural or faunal diversity. It once housed red wolves, Florida panthers (*Felis concolor coryii*), and ivory-billed woodpeckers. Nonetheless, the reserve is still home to more than four hundred species of mammals, birds, reptiles, amphibians, and fish, including the largest number of endangered Louisiana black bears in northern Louisiana. As many as two hundred bears may reside in the Tensas area. If refuge managers implement new procedures to manage trees for old growth, including critical dead and dying trees, there is a chance that ivory-bills may wing their way back to Tensas (U.S. Fish and Wildlife Service, Tensas River National Wildlife Refuge Profile).

The attention the rediscovery of the ivory-bill in Arkansas has generated will likely place pressure on management personnel to reevaluate forest-management strategies in Tensas. However, changing how the Tensas forests are managed may be no easy task. The Tensas River

National Wildlife Refuge is also managed for game animals such as deer. Deer hunting is extremely popular in Louisiana. Some cutting and thinning of timber takes place in Tensas to improve deer habitat, and therein lies the potential conflict between the needs of wildlife (like the ivory-bill) that prefers large trees and dense forests and the needs of deer, which prefer more open forest. This potential conflict harkens back to the observation by Dwight LeBlanc that was discussed in chapter 3: One of the biggest challenges to the ivory-bill's recovery will be how local people perceive the effects of forest management and related conservation issues on their lifestyles. It may be, however, that the forestry and wildlife managers at Tensas will determine that simply leaving some old, especially dead and dying trees in place may be enough to support a small ivory-bill population since the ivory-bill appears to be more flexible in its habitat requirements than previously thought.

Beyond Louisiana: East Texas

In addition to Louisiana, the Big Thicket region in east Texas should be placed on the primary list of search locations. And, in fact, in March 2006 the first call went out from the Gulf Coast Bird Observatory to the birding and ornithological communities for volunteers to search the Big Thicket area. This area was made famous within ivory-bill and birding circles by John Dennis's reports of seeing and hearing ivory-bills in this region in the late 1960s. Dennis believed there were probably five to six pairs of ivory-bills still alive in the Big Thicket in the late 1960s. And Dennis, like George Lowery, came under fire for these reports, with some in the ornithological community questioning his reputation and honesty. For example, James Tanner dismissed Dennis's reports outright, as he had dismissed many other ivory-bill reports over the years. The Arkansas sightings, however, have changed the way other sightings, both past and present, should be evaluated. The Cornell Laboratory of Ornithology recently reviewed some of Dennis's old and controversial ivory-bill recordings from Texas, and while the recordings were not conclusive, the drumming and calling were not dismissed either: They could have been ivory-bills.

The Big Thicket region is similar to the White River area in Arkansas where the ivory-bill was filmed, and much of it is managed by the federal government as a designated national preserve. If an ivory-bill is found in the area, the fact that the area is federally protected will make it easier to implement a recovery plan. The preserve consists of nine land units and six river corridors encompassing more than 97,000 acres (U.S. National Park Service, Big Thicket National Preserve Profile). Not all of this is suitable ivory-bill habitat and not all of it is contiguous, but certainly much of the bottomland forest is potential habitat. Tanner visited the area in the late 1930s and stated the forests were totally inappropriate because they were cut over, but in the last seventy to eighty years, much of the bottomland forest has regrown and become government-protected land. Like the Atchafalaya Basin, the Big Thicket area has been greatly impacted by humans, but it contains many wild spaces, wild enough that cougars have reappeared in the reserve. Also similar to some areas in Louisiana, many solid reports of ivory-bills have come from the Big Thicket area during the latter half of the twentieth century. And, as Chuck Hunter of the U.S. Fish and Wildlife Service told me, "Where there is smoke, there's fire."

Florida

Many solid reports of sightings of the ivory-billed woodpecker have come from Florida (Stevenson and Anderson 1994). One must wonder what might have happened if the conservation community had put as many resources into searching the Chipola River region in the 1950s as it put into searching for the ivory-bill in Arkansas. Perhaps if the conservation community had invested more resources in Florida, the term "rediscovery" would be absent from our discussion of the ivory-bill today.

A few areas in Florida stand out in terms of sightings and should be added to the list of places to search. First, the areas that encompass the Suwannee, Apalachicola, Choctawhatchee, and Lower Chipola rivers in the Florida Panhandle should be given some attention (and the Choctawhatchee River is being searched as this volume goes to press). While this list describes a diverse set of rivers and ecosystems, areas along these waterways with mature bottomland forests should at least be surveyed regarding forest, maturity, and species composition. Many areas along these Panhandle rivers are similar to other locations in the South that house good ivory-bill habitat because they receive some sort of federal or state protection. For example, the Apalachicola National Forest, containing over 564,000 acres, is one of the largest contiguous blocks of public land east of the Mississippi River (U.S. Dept. of Agriculture, Apalachicola National Forest Profile). The area contains a variety of ecosystems such as longleaf pine, savanna, and bottomland forests, and—in the southern end of the national forest—a bay ecosystem. The area is also home to the world's largest and most viable population of the endangered red-cockaded woodpecker, as well as having one of the state's largest populations of black bears.

Although most of this terrain in the Apalachicola National Forest is not appropriate ivory-bill habitat, the vast reserve protects great ecological diversity. Searches should concentrate on the areas dominated by bottomland forests—especially in the southern Chipola River and northern Apalachicola River areas, which have generated solid reports of ivory-bill sightings.

In addition to bottomland forests in the Florida Panhandle, the Big Cypress region offers another promising area in which to search for ivory-bills, especially the 100,000-acre Fakahatchee Strand. Big Cypress National Preserve, like its counterpart in the Panhandle, has a mixture of pines, hardwoods, prairies, mangrove forests, and cypress stands and domes. White-tailed deer, black bear, and the Florida panther can be found in Big Cypress. This area is an ecotone where both temperate and tropical species mingle to create one of the most biologically diverse landscapes in North America (U.S. National Park Service, Big Cypress National Preserve Profile). It has been many years since ivory-bills were reported in the Big Cypress region, but sporadic rumors have filtered in over the years (Jackson 2004).

Solid reports of ivory-bills over the years from other areas of Florida are quite sporadic, not indicating clear geographic patterns. There have been a few reports from central Florida and a few reports from northern Florida. Following up on these random reports could be time-consuming and costly since there is no a centralized region or protected area upon which search efforts would do well to focus.

South Carolina

According to Chuck Hunter, a couple of areas in South Carolina are often mentioned as having potential for more sightings of the ivory-bill. Although fewer recent reports have come from South Carolina than from Louisiana, good ivory-bill habitat does exist there.

At the top of this list is Congaree National Park. Designated in 2003, it is one of the most recent national parks (U.S. National Park Service, Congaree National Park Profile). Congaree protects the largest of the few remaining tracts of old-growth cypress and bottomland forests left in North America. About ninety tree species are found among the 11,000 acres of bottomland forest. Congaree contains a great deal of old, dead, and naturally dying timber, meaning that the forest is evolving through a natural cycle of life and death, creating ideal conditions for the ivory-bill (Jones 1997; B. P. Allen and Sharitz 1999). Congaree contains enough old-growth habitat that it is also rumored to possibly house the extinct or nearly extinct Bachman's warbler (*Vermivora bachmanii*) (Watson and Koches 2002). John Dennis reported that during his stay in the swamp, he received many accurate descriptions of ivory-bills from hunters and fishermen who frequented the swamp and claimed to have sighted the bird (Dennis 1988).

The Congaree area also has an interesting human as well as ecological history. General Francis Marion, aka the Swamp Fox, used the swamp as a hideout during the Revolutionary War between battles with the British. Marion, who depended on the shelter the great trees and swamp provided, wrote in his journal: "I look at the venerable trees around me and know that I must not dishonor them" (U.S. National Park Service, Congaree National Park Profile). While the swamp was from time to time targeted for extensive logging, its inaccessibility kept it from being completely razed. Now that it is forever protected as part of the national park system, the trees will continue to be honored as Marion desired.

Moving toward the coast and away from Congaree National Park, one enters the Lower Santee River region and the Francis Marion National Forest. This region of bottomland and low coastal forests has provided many historical reports of ivory-bills, especially near the coast. The Francis Marion National Forest is a vast forested and swampy region. Like the other landscapes described in this chapter, the national forest houses a great variety of plants and animals, many rare and endangered. Hurricane Hugo heavily damaged many of the mature trees in Francis Marion in 1989. However, if ivory-bills are present there, this damage could actually benefit them because the damaged and dead trees might offer the birds better food and nesting trees. Also, that the best ivory-bill habitat in South Carolina has federal protection/management status is a cause for guarded optimism.

Other States

Based on past reports and current habitat, a handful of other locations might be appropriate for ivory-bill searches. While reports from these areas might be quite old or sporadic, the fact that some of these areas now house mature second-growth bottomland forest means there is a slight chance that ivory-bills might be present.

In Mississippi, the Pascagoula Swamp in the southeastern part of the state has provided solid reports of ivory-bills over the years. Similarly, the Yazoo Delta near the Mississippi River in west central Mississippi has yielded reports, including one in 1987 by woodpecker expert Jerome Jackson (Jackson 2004). In addition to these two areas, the fairly large forest patches along the Mississippi River in Mississippi and Louisiana might provide some habitat or at least a corridor for the movement of ivory-bills. These batture forests may have been critically important for ivory-bills as they left Tensas in the 1940s and moved into Arkansas, Mississippi, or southern Louisiana. This, of course, is speculation; we do not yet know the true ecological importance of these forest corridors.

In Alabama, the Mobile Delta swamp is also an extensive and maturing forest. When I visited the area in 2005, I was impressed by the size of many of the trees. In Georgia, the Lower Altamaha River has some good bottomland forest habitat, as does the huge Okefenokee Swamp and Savannah River Swamp along its border with South Carolina. Tanner commented at the 1986 ivory-bill meeting that the Altamaha River contained some of the best ivory-bill habitat he had seen. And even extreme western Tennessee contains some good habitat near the Mississippi River. An unnamed, but apparently fairly well-known ornithologist reported seeing an ivory-bill in western Tennessee in 2006. Few details were available at the time of this writing. Most of the areas that have supplied ivory-bill reports during the twentieth century contain mature bottomland forest habitat today.

While my listing of all these areas as having potential for housing ivory-bills may seem overly optimistic, it is important to remember that before the Arkansas sighting, the White River region was not at the top of most experts' lists for places to search for the bird. The fact that it was not listed as a priority search area by the initial ivory-bill search committee, which was made up of the foremost woodpecker experts, is a sober reminder of just how little was understood about the bird and its habitat.

AFTERWORD

A book of this nature—a story of the rediscovery and the continuing search for an extremely rare and reclusive species—really can have no conclusion. There were and are many more stories to record and sightings to track down. If ten or even one hundred more ivory-bills are found, the story doesn't end there. The recovery and management phase, perhaps the trickiest part of the story, will just then begin. Much more will be written about the ivory-billed woodpecker. The debate about forest management, conservation biology, and the authenticity of the Arkansas sightings and film has already begun.

And while other stories about the ivory-bill remain to be told—and many will come from areas outside of Louisiana, such as east Texas and South Carolina—I end this book where it began, on a final trip to the southern edge of the Atchafalaya Basin. This region, especially the Franklin-Patterson area, was the source not only of the best and most colorful ivory-bill stories but also of some of the most recent sightings. Based on the reports of the 2005–6 Louisiana search team, I am confident that, had Hurricane Katrina not disrupted the search, there would now exist incontrovertible documentation of the ivory-bill. This is not just an expression of my own wishful thinking; Keith Ouchley, organizer of the search, among many members of the search team, believed that another rediscovery announcement was imminent before Hurricane Katrina returned the search for the ivory-bill in Louisiana to square one.

So once again I find myself driving along the eastern and southern edge of the Atchafalaya Basin as I did many times during the years I researched this book. I have always loved this drive with the marsh, swamp, and bottomland forest on both sides of the road. In the back of my mind, as always, is the thought that a large woodpecker may fly across the road—you never know. And various waterfowl and a pileated woodpecker do fly over the road, but no Lord God Bird. The drive through this unique landscape—with the towering cypress along the road draped with Spanish moss and with signs tacked to the trees offering goo, gar, and crawfish for sale—and the conversations I have had with the people who live near the ivory-bill's past and present haunts always remotivated me when the writing slowed to a trickle.

As I drive down and around the edge of the Atchafalaya Basin, it strikes me that there are still many wild spaces in this part of the state. Yes, the area's forests have been logged, and yes, the forests and bayous are visited by fishermen and hunters. But today, as I pass camps, bayous, and houseboats, it somehow seems easy to imagine an ivory-bill in some of this mature forest and swamp. And when I look out over the vast area with its junglelike vegetation, I think, once again, that it is a small wonder that the ivory-bill has not yet been photographed. Until one actually travels into a bottomland forested area, it is hard to appreciate the complexity, size, and challenges that a place like the Atchafalaya Basin presents to those searching for a small number of extremely wary birds.

When I made my first trip to Franklin in 2003, I was not sure what lie in store regarding this book or the ivory-bill. I certainly had no inkling that the ivory-bill would make national and international headlines in May 2005. Even though a few people, such as Chuck Hunter, hinted that there might be a few ivory-bills in Arkansas, most experts had followed Tanner's lead and written off the former White River area as potential ivory-bill habitat because so many hunters and fishermen visit it and because, like most of the bottomland forests in the South, so much of it had been cut over. When I made this drive for the first time in 2003, it was to meet with Garrie Landry in Franklin. Garrie would put me in touch with many individuals who had ivory-bill experiences, including Fielding Lewis, who led me to the area resident who had most recently seen ivory-bills, Scott Ramsey. The story of one person's ivory-bill sighting always led me to others. And as I followed the trail of these stories, it seemed as if I could continue recording and writing them forever, that they had no end. What this web of stories bespoke, I came to realize, was a deep connection and fascination with the ivory-bill among the people who best know the countryside of rural Louisiana.

When I started the project in 2003, a few people I told about it laughed and suggested it was similar to writing a book about Bigfoot. And the first time I visited the Louisiana Department of Wildlife and

Fisheries office in Baton Rouge, few people took the book or, more important, the larger issue very seriously. The typical response from many in the office, other than Nancy Higginbotham, seemed to be, "Don't you have anything better to do?" While I never took these skeptics and their comments personally, I can't help but feel vindicated by the recent searches and sightings in Arkansas and Louisiana. The ivory-bill is no longer a Bigfoot.

The story of the ivory-bill's rediscovery decades after it was pronounced gone forever is both remarkable and inspiring. And the fact that the ivory-bill's habitat has been severely disturbed by human actions makes the bird's survival even more amazing. I hope this event sets in motion a reappraisal of how we manage our bottomland forests. Instead of managing primarily for deer, game birds, and timber production, we should begin to increase the management of our forests for more complex climax communities. Nongame species such as songbirds are beginning to show their potential economic value. The development of birding trails in various states, including Louisiana, is partly a response to the dawning recognition of the increasing economic value of nongame species. Although today's ecotourism industry may be most developed in the tropical world, it is also a rapidly expanding economy in places such as Louisiana where a great deal of diversity still exists.

Another important and developing issue is what the recent damage caused by the hurricanes in Louisiana will mean for wetland and bottomland forest conservation. Will federal and state authorities invest more resources in these areas and manage them differently because of their ability to absorb floodwater? Will reforestation and general conservation become more prominent state and federal activities, especially in the southern sections of Louisiana? One good that might come of the tragic events of 2005 would be that we begin to consider the marshes and bottomland forests to be true national treasures.

Saying that the time is right for greater emphasis on the management of forests for nongame species and floodwater control, however, should not mean that hunters and fishermen will be excluded from any future plans involving ivory-bills. Ivory-bills survive today because the hunters and their clubs have set aside large tracts of bottomland forests, and their dollars help support both federal- and state-protected lands such as the Cache River and White River National Wildlife Refuges. Many of the best reports of ivory-bills, especially in the recent past, have been made by avid hunters like David Kulivan, Jay Boe, and Scott Ramsey—not by birders. As I have argued, ivory-bill recovery and conservation is compatible with the activities of hunters and fishermen.

Some forestry practices can and should change—girdling more trees, leaving more old timber, and managing for climax communities in some areas. Developing a larger "landscape" approach in the management of our bottomland forests will certainly be beneficial to more than just ivory-bills; other plants and animals, including many game animals will also benefit from new conservation policies. This habitat compatibility between game and nongame species is one reason that Ducks Unlimited—one of the largest and most powerful hunting and conservation groups in North America—is tying some of its new initiatives such as reforestation efforts in the White River region in Arkansas to the rediscovery of the ivory-bill. Ducks Unlimited is one of the best examples of a private sporting organization that recognizes the value of and has worked to restore bottomland hardwood forests that provide habitat for millions of migrating and wintering waterfowl every year, especially in the Lower Mississippi River alluvial valley. According to Phil Covington, a Ducks Unlimited biologist, "We should really thank the ivory-billed woodpecker . . . for bringing bottomland hardwood forests to the forefront of public conservation priorities" (Ducks Unlimited 2005:2).

The rediscovery of the ivory-bill may also set in motion a new appreciation not only for this species and its habitat but also for conservation in the South in general. In conversation after conversation for this book, individuals told me how the ivory-bill is a symbol of the destruction of the southern environment. The rediscovery of the ivory-bill could offer a watershed moment that drives public interest and support for conservation issues. The ivory-bill almost slipped away altogether because of the shortsightedness of individuals like the executives of the Chicago Mill and Lumber Company, which owned the timber rights to the former Singer Tract in the 1940s. We now have a chance to make amends for past mistakes by doing everything possible to ensure that the ivory-bill and its habitat survive so that future generations never have to wonder if the ivory-bill is extinct.

I stop in my drive through the Louisiana countryside for a long, quiet look at the Atchafalaya Basin. I climb to the top of the levee in an area that provides an expansive view from above. Casting back to the beginning of this journey and project, I remember that I was never sure then whether I would weigh in definitively on either side of the controversy about whether the bird exists. As it turns out, the ivory-bill resolved that dilemma for me. There is little doubt that somewhere, in that riotous assemblage of trees, vines, and shrubs, ivory-bills still fly.

Appendix 1
Locations of Past Ivory-Bill Sightings

Sightings are mapped and listed by decades starting in the 1950s because it was documented by James Tanner that the ivory-bill survived into the 1940s in Louisiana (map 4). Tanner believed, as did most other ornithologists at the time, that the ivory-bill went extinct after the former Singer Tract (now the Tensas River National Wildlife Refuge) was logged in the early 1940s (Tanner 1942a). Thus sightings after the 1940s are most relevant to the current debate concerning the present status of the ivory-bill.

1950s

Several credible sightings were reported in the 1950s. While there was no center of concentration for these reports, three areas stand out where future sightings would also be made: the Pearl River region in Louisiana; the Chipola River in the Florida Panhandle; and South Carolina's Francis Marion National Forest (Dennis 1988; Jackson 1989). The Chipola River was the location of searches and sightings of the ivory-bill in the early 1950s by naturalist Whitney Eastman (Baker 1950; Dennis 1967; Eastman 1958). His reports drew enough attention that an area was set aside as an ivory-bill sanctuary for a short period, but it was later abandoned after the birds were neither seen nor heard in the area (see Eastman 1949, 1958).

1960s

The 1960s witnessed a number of sightings, some from new areas as compared with the 1950s, and some in similar or identical areas. Most interestingly, the Big Thicket region in east Texas produced a number of reports of ivory-bills (see the works by John Dennis in the reference list). Naturalist-author John Dennis was convinced, based on his own and others' sightings, that east Texas housed several pairs of ivory-bills into the late 1960s. The Florida Panhandle produced a few more sightings, as did southwest Florida near Lake Okeechobee. Coastal South Carolina and Francis Marion National Forest also continued to produce ivory-bill sightings.

1970s

The 1970s saw a flurry of sightings from the Atchafalaya Basin. The most famous of these was made in the far southern edge of the basin near Franklin, Louisiana (map 1). The sighting's photographer, Fielding Lewis, sent some of the photos to ornithologist George Lowery at LSU. Lowery later visited the area with Lewis but failed to see any ivory-bills. Fielding Lewis, whose identity was kept secret by Lowery, saw ivory-bills on several occasions in the 1970s and 1980s (Lowery 1974; Lewis 1988). East Texas also produced two more sightings, lending further support to John Dennis's earlier claims.

1980s

Louisiana was again the location for most ivory-bill sightings in the 1980s. The Atchafalaya Basin and the Pearl River area accounted for most of the sightings. Several of these reports were made by wildlife and conservation professionals, individuals who in interviews claimed to know exactly what they were looking at—ivory-billed woodpeckers. For example, Nancy Higginbotham, once employed at the Louisiana Department of Wildlife and Fisheries in the Non-Game Division, claims to have seen the ivory-bill in the Pearl River area twice—a male on the edge of the Pearl River swamp in 1986 and a female while she was deer hunting in the Pearl River Management Area in 1987. Higginbotham is an experienced outdoorsperson who grew up in the area and knows its flora and fauna. In addition to Louisiana, east Texas was the location for several more ivory-bill sightings in the 1980s.

1990s

The 1990s proved to be a quiet decade for ivory-bill sightings. However, David Kulivan's 1999 report from the Pearl River Wildlife Management Area provided the impetus for the 2002 search in the Pearl River area that aroused national interest in the ivory-bill.

Post-2000

The 2002 search in the Pearl River Wildlife Management Area created a media frenzy with coverage appearing on radio, television, and newspapers throughout the United States. It also rekindled interest in finding the ivory-bill among many in the ornithological and birding communities. New reports began to be taken more seriously by some in the ornithological community, and as a result, sightings trickled in from various areas in the South. Sightings from Arkansas and the Atchafalaya Basin region in Louisiana turned out to be the best leads. In addition to the several reports coming from the Patterson area described above, the ivory-bill also was reportedly seen in 2001 by an LSU graduate student driving on Route 83, near Franklin, and by a hunter near the Tunica Hills north of Angola State Penitentiary on the banks of the Mississippi River. One of the most recent and best sightings in Louisiana was made on September 7, 2005, near Bayou Sorrell in the Atchafalaya Basin. An ivory-bill was seen flying above the canopy by Ron Boustany, a natural resources specialist with the U.S. Department of Agriculture, while flying over the Atchafalaya Basin in a small airplane during a Hurricane Katrina damage survey. A follow-up ground survey in the area did not report any sighting of the bird; however, extensive tree scaling was reported. This region within the Atchafalaya Basin has produced many sightings over the past decades. In Florida, the area around the Choctawhatchee River produced a new round of ivory-bill sightings and news stories in 2006 (Hill et al. 2006). A team of ornithologists led by Auburn University professor Geoffrey Hill will expand their search in an effort to finally get the elusive, indisputable photograph or video recording of the bird. Although their search has apparently recorded ivory-bill drumming and calling and has had numerous visual sightings, they have not produced a photograph. And without a clear photograph, even the best audio recordings will continue to be criticized by many skeptics. As of this time, the search continues.

Appendix 2

Timeline of Reported Ivory-Bill Sightings in the United States

The following timeline is a collection of reported sightings from roughly the past ninety years (see map 4). Some of the reports were taken from published sources such as Jerome Jackson's U.S. Fish and Wildlife report (1989), as well as Internet resources such as *Birder's World Magazine* and the Nature Conservancy. However, I have also included many previously unknown sightings in Louisiana that I have come across in the files of the U.S. Fish and Wildlife Service and in the personal files of Tommy Michot. Also listed are other sightings relayed to me by various individuals who were interviewed for this book. While none of these reports have accompanying evidence such as photos (except the 1971 Fielding Lewis report), they should be considered as potential leads in identifying areas in which to conduct future searches. I have included in this timeline only reports made by experienced birders or naturalists or by individuals whose accounts provided enough very specific detail to warrant serious consideration.

April 13, 1924: Arthur Allen photographed a pair of ivory-billed woodpeckers in central Florida. The birds were later collected.

1920s: Individual birds were seen near Deer Park, in Osceola County, Florida.

April 1932: Mason Spencer shot a male ivory-bill in Madison Parish, Louisiana, around the Tensas River. The bird's presence there surprised the Louisiana fish and game authorities, which drew the attention of the ornithological community and led to the work of Allen, Allen and Kellogg, and later Tanner in the Singer Tract (now the Tensas River National Wildlife Refuge) in the late 1930s.

1933: LSU ornithologist George Lowery saw a pair of ivory-bills in the Singer Tract on Christmas Day.

1934–35: Ivory-bills were regularly seen along the Lower Santee River, South Carolina.

1937–39: Tanner completed his dissertation work on the ivory-billed woodpecker. A few pairs were present in the Singer Tract.

August 1941: Three ivory-billed woodpeckers were seen in the John's Bayou area of the Singer Tract by George Bick and Jim Parker.

December 1941: Tanner found an adult and juvenile female in a cut-over area of the Singer Tract.

May 1942: Roger Tory Peterson and Bayard Christy observed two females in the same area of the Singer Tract where Tanner saw ivory-bills in 1941.

November 1942: John Baker identified a single female in the Singer Tract.

1941–42: Two ivory-bill sightings were made in the Okefenokee Swamp, Georgia.

Early 1940s: Several ivory-bill sightings were reported in Bolivar County, Mississippi, sometimes of more than one bird.

1943–44: Richard Pough of the National Audubon Society sighted a single female in the Singer Tract.

1944: Don Eckelberry saw and sketched a female in the Singer Tract. For a time, this was believed to be the last ivory-bill sighting in the Singer Tract.

1949: Reports in southeast Missouri of an ivory-bill. This is the last report from Missouri.

Late 1940s: Deer hunters saw several pairs of ivory-bills south of Tallulah, Louisiana.

1950: Sighting of an ivory-bill in the Big Cypress Swamp in Florida.

1950–52: Whitney Eastman searched for and reported finding ivory-billed woodpeckers along the Chipola River in Florida.

April 1951: John Dennis reported hearing an ivory-bill along the Chipola River in Florida.

1952: Sam Grimes and Roy Hallman reported seeing an ivory-bill in Wakulla County, Florida.

1955: John K. Terres reported seeing an ivory-bill south of Homosassa Springs, Florida.

1955: A male ivory-bill was seen and heard on the east side of the Pearl River in Mississippi near Lock Number 1.

1956: A pair of ivory-bills was seen in an eastern North Carolina bottomland cypress swamp.

1958: An ivory-bill was seen from a small plane flying over the canopy in the Altamaha River Basin, Georgia.

1958: A pair of ivory-bills was reported near Thomasville, Georgia.

1959: William Rhein reported seeing an ivory-bill west of the Aucilla River in Florida.

1959: An ivory-bill was seen and heard in the Francis Marion National Forest near Awendaw, South Carolina.

1959: A game warden saw an ivory-bill near Awendaw, South Carolina, near Moore's Landing area.

December 1960: Two ivory-bills were seen in Leaf River Swamp, near U.S. Highway 98 in Perry County, Mississippi.

1960: An ivory-bill was seen on numerous occasions on Bull's Island, South Carolina.

1962: An ivory-bill was seen on Highway A1A, 3 miles north of Indialantic, Mississippi.

1962: An ivory-bill was reported from the area west of Summerville, South Carolina.

1963: An ivory-bill was seen in the Francis Marion National Forest, South Carolina, on Iron Swamp Road.

1963: "Clear" sighting of an ivory-bill in southwest Georgia, near the Georgia-Florida border.

Late 1963: An ivory-bill was seen flying across road about 10 miles west of Apalachicola, Florida.

1965: An ivory-bill was seen in the Okefenokee Swamp in Georgia.

1966: An ivory-bill was seen in the Neches River swamp near Steinhagen Reservoir in Texas.

1966: Two ivory-bills were seen feeding in dead pines near Eglin Air Force Base in the Florida Panhandle.

1966: John Dennis reported seeing an ivory-bill in the Neches River swamp in Texas.

1966: John Dennis reported hearing a typical ivory-bill call in the Neches River swamp in Texas.

1967: Two ivory-bills were seen in the Francis Marion National Forest near McClellanville in the Buck Hall area, South Carolina.

1967, 1969: A pair of ivory-bills was seen in 1967 on a ranch in Polk County, Florida, northwest of Lake Okeechobee. A single bird was seen several times through 1969.

1968: A game management officer saw an ivory-bill in the Navasota River bottom, Texas.

1968: An ivory-bill was seen on large rotting log in Sabine River valley, Texas.

1968: John Dennis recorded what he believed was an ivory-bill woodpecker in the Neches River swamp near Village Creek in Texas.

Fall 1968: An ivory-bill was reported from Highland Hammocks State Park, near Sebring in south central Florida.

1969: The double-knock hammer of the ivory-bill was heard in the Big Thicket area of east Texas.

Early 1970s: James "Butch" Huff observed ivory-bills on several occasions while boating in the Atchafalaya Basin (on the Atchafalaya Cut near Grand River and Butte la Rose).

1971: Fielding Lewis photographed an ivory-bill on different tree trunks in the far southern Atchafalaya Basin, near Franklin, Louisiana.

1971: An ivory-bill was spotted near Patterson, Louisiana, in the southern Atchafalaya Basin.

1971: Forest rangers saw an ivory-bill in the Sam Houston National Forest in Texas.

1972: Robert Hamilton observed an ivory-bill flying over I-10 west of Baton Rouge near Ramah, Louisiana.

March 1973: Jerome Jackson reported seeing an ivory-bill while canoeing on the Noxubee River, Mississippi and Alabama.

1973: A father and son reported seeing an ivory-bill on the Ogeechee River, 25 miles west of Savannah, Georgia.

1973: A fisherman saw an ivory-bill near Duck Lake in the southern Atchafalaya in Louisiana.

1974: A fisherman saw an ivory-bill near Duck Lake in the southern Atchafalaya in Louisiana.

1974: Sighting of ivory-bill near Upper Grand River in Atchafalaya Basin, Louisiana.

1974: Robert Bean saw an ivory-bill flying across Interstate 10 in the Atchafalaya Basin about 32 km west of Baton Rouge, Louisiana.

1976: A male ivory-bill was seen in flight near the Steinhagen Reservoir, Neches River, Texas; another bird or possibly the same bird was seen several miles away the next day.

1976: University of the Wilderness staff observed an ivory-bill as it flew across a road near Steinhagen Reservoir in the Big Thicket area of east Texas.

1978: Two biologists heard the call of an ivory-billed woodpecker in Black Creek, DeSoto National Forest, Mississippi.

1978: Robert Hamilton heard an ivory-bill call and double-rap near Grand River in the eastern Atchafalaya Basin in Louisiana.

1978: LSU graduate students heard an ivory-bill call in the Grand River area in Louisiana, near the location where Hamilton heard a bird.

1978: An ivory-bill was spotted from a helicopter flying over the eastern Atchafalaya Basin in Louisiana.

1978: A biologist reported sighting a female ivory-bill while canoeing on the Amite River near the Mississippi and Louisiana state line.

Late 1970s–early 1980s: An ivory-bill was reported in Sicily Island, Louisiana.

1981: A pair of ivory-bills were sighted in the Big Thicket National Preserve in east Texas by a preserve volunteer.

1981: Tommy Michot and David Hankla observed a possible ivory-bill near Duck Lake in the southern Atchafalaya in Louisiana.

1981: Heinrich and Welch heard an ivory-bill call in the former Singer Tract area in Louisiana.

1982: An ivory-bill was sighted in western North Carolina near Connestee Falls, with black crest, single-note call, and white on "downturned wing."

1982: An ivory-bill was seen in Pearl River forest in Louisiana.

1983: A father and son fishing saw a black-crested woodpecker, flying straight in an eastern North Carolina bottomland cypress swamp.

1984: An ivory-bill was seen flying at treetop level near Wateree River in South Carolina.

1985: An ivory-bill was seen feeding in a dead palm along the Loxahatchee River in Dickinson State Park in Florida.

1985: An ivory-bill was seen in Paradis Canal in St. Charles Parish, Louisiana.

1985: An ivory-bill was seen on the eastern side of the Toledo Bend Reservoir in Louisiana near Turtle Beach Marina.

1985: An ivory-bill was seen on the western side of the Toledo Bend Reservoir in Texas.

1985: An ivory-bill was reported near Chireno, Texas, in Nacogdoches County.

1986: Nancy Higginbotham and her mother observed a male ivory-bill on edge of Pearl River swamp on LA 41 in Louisiana.

1986: A fisherman and his son saw a pair of ivory-bills in the Grand River area in Atchafalaya Basin in Louisiana.

1986: Edwin Broussard, enforcement agent with Louisiana Wildlife and Fisheries, saw a male ivory-bill near Bayou Long, 6 miles north of Duck Lake in the Atchafalaya Basin in Louisiana.

1987: Nancy Higginbotham observed a female ivory-bill for several hours while deer hunting in the Pearl River swamp in Louisiana.

1987: Jerome Jackson and graduate student heard an ivory-bill near Vicksburg, Mississippi.

1988: The final year that Fielding Lewis saw an ivory-bill near Franklin, Louisiana.

1988, 1989: An ivory-bill was seen on two occasions on private land in St. Helena Parish, Louisiana.

1990: An ivory-bill was seen flying near Butte LaRose in the Atchafalaya Basin.

1993: An ivory-bill seen working a dead stump near Bayou Sorrel, on the edge of the Atchafalaya Basin in Louisiana.

April 1, 1999: David Kulivan reported seeing two ivory-bills in the Pearl River Wildlife Management Area.

2001: A male ivory-bill was seen flying straight and at treetop level near the Apalachicola National Forest, Florida.

2001: An ivory-bill was seen by an LSU graduate student driving on Route 83, near Franklin, Louisiana.

2002: An ivory-bill was spotted by a hunter near the Tunica Hills, north of Angola State Penitentiary in Louisiana.

2002: A fisherman saw a pair of ivory-bills west of Bayou Sorrel in the Atchafalaya Basin, Louisiana.

September 7, 2005: An ivory-bill was seen by a USDA biologist flying over Atchafalaya Basin near Bayou Sorrel in Louisiana.

2004–5: An ivory-bill was seen and videotaped in the Cache River region, Arkansas.

2005: An ivory-bill was seen and heard on several occasions on private land near Patterson, Louisiana.

2005–6: Ivory-bills were seen and heard on multiple occasions near the Choctawhatchee River in the Florida Panhandle by Auburn University ornithologist Geoffrey Hill and graduate students.

2006: Ivory-bill reported from western Tennessee near Mississippi River.

2006: Jay Huner saw an ivory-bill near Patterson, Louisiana, in June 2006.

2007: Scott Ramsey saw and heard ivory-bills on several occasions at his hunting camp near Patterson, Louisiana, in the first half of 2007.

Bibliographic Essay

Publications on the ivory-billed woodpecker have found a variety of outlets, including peer-reviewed scientific journals, popular magazines, and books. While the reference list contains many major and minor publications on the ivory-bill, bottomland forests, and specific protected areas suggested as sites for future searches—including all of the works I cite in this volume—I would like to briefly mention here some of the more important works that I consulted while researching this book.

Several books have been published that provide extensive information about the ivory-bill's natural history, past searches, famous searchers, and factors responsible for the bird's decline. James Tanner's classic publication *The Ivory-billed Woodpecker* provides much of the information on which we base our opinions about ivory-bill behavior. This book is the result of Tanner's dissertation fieldwork in Madison Parish, Louisiana, in the late 1930s. While assumptions made in this work are now being carefully analyzed in light of recent events, Tanner's work remains an important, and the most extensive, ecological study of the ivory-bill. More recently, Jerome Jackson's *In Search of the Ivory-billed Woodpecker* (2004) provides an important update and expansion of Tanner's earlier work. Jackson covers an array of topics, including past searches for the bird, habitat assessment, the ivory-bill in Cuba, and the bird in popular culture. Jackson's book provided invaluable information for my work on topics such as Native American uses of the ivory-bill and the distribution and sightings of the bird in the early twentieth century. Tim Gallagher's 2005 *Grail Bird: Hot on the Trail of the Ivory-billed Woodpecker* provides a behind-the-scenes description of the Arkansas search and rediscovery from start to finish. Gallagher discusses past searches and sightings and some of the major figures in the search for the bird. Philip Hoose's *The Race to Save the Lord God Bird* (2004), while intended for a younger audience, offers useful information about the factors that drove the ivory-bill to the brink of extinction.

Numerous articles on the ivory-bill have appeared in both popular publications as well as scientific journals. Most publications prior to the 2004 rediscovery debated the ivory-bill's existence or reported on sightings in both the United States and Cuba. For instance, there are many articles that describe sightings of the ivory-bill in the Florida Panhandle in the 1950s, and later in east Texas. Post-2005 publications focus on the rediscovery, on whether there truly was a rediscovery, and on plans for saving the ivory-bill if it still exists. Noteworthy in this group are the original 2005 article in *Science* by John Fitzpatrick et al.; the rebuttal to the rediscovery news by birder and artist David Sibley, also in *Science* (Sibley et al. 2005); Jerome Jackson's 2006 critique, in the *Auk*, of the evidence provided by the Cornell search team (Jackson 2006); and the search team's response to Jackson's comments, also in the *Auk* (Fitzpatrick et al. 2006). Other important publications include those by ornithologist Lester Short that describe his searches for the ivory-bill (Short 1985, 1987, 1988; Short and Horne 1986), as well as those by John Dennis, who, while derided by James Tanner, provided invaluable information about the ivory-bill's status in northern Florida and east Texas (Dennis 1948, 1967, 1979, 1984, 1988). Tanner dismissed Dennis's reports of ivory-bills in the 1960s and 1970s, and called his recording of an ivory-bill call a fake. However, the recording has since been analyzed at Cornell, with the results being inconclusive. Jerome Jackson has also written a number of useful articles over the past couple of decades while he was searching for the ivory-bill and compiling information for his book (Jackson 1989, 1991, 1996, 2002a, 2002b).

There are also many older and somewhat obscure publications focused on the ivory-bill. Although dated, some provide interesting snippets of information about past ivory-bill distribution, behavior, and human impacts. Several prominent ornithologists studied and wrote about the ivory-bill, including Arthur Allen, the founder of the Cornell Laboratory of Ornithology and the individual who took the first photographs of a living pair of ivory-bills in 1924 (Allen 1924, 1937, 1939; Allen and Kellogg 1937), and George Lowery at LSU. In his *Louisiana Birds* (1974), Lowery provides a moving description of his sighting of an ivory-bill on Christmas Day 1933 in the Tensas River area. James Tanner, in addition to his published dissertation, wrote two articles that were summations of his larger work (Tanner 1941, 1942b).

There are also several publications that provide evidence about the former range of the ivory-bill, as well as its importance to Native American cultures in North America. These studies are interesting because they point out two important facts about the ivory-bill: first, that it previously existed across the South as well as into the central Mississippi River (possibly to St. Louis) and Ohio River basins; and second, that Native Americans far beyond the ivory-bill's natural range valued its plumage and bills. Both Audubon and Catesby provide interesting descriptions of the Native American use of ivory-bill parts (Audubon 1834; Catesby 1985). Jerome Jackson provides a complete chapter on the issue of Native Americans and ivory-bills in his 2004 book.

In addition to publications focused exclusively on the ivory-bill, there are many books describing the southern landscape and its people that contributed to my understanding of the territory that the ivory-bill once inhabited. The works by Audubon (1834, 1842, 1999) and Catesby (1985) are largely focused on bird life and wildlife, while others explore the cultural landscape. James Cobb's 1994 *The Most Southern Place on Earth: The Mississippi Delta and the Roots of Regional Identity* is a classic work for those interested in better understanding the Delta region, and Ted Ownby's *Subduing Satan: Religion, Recreation and Manhood in the Rural South* (1990) examines hunting culture in the South before a conservation ethic developed. William Faulkner's *The Bear* (1961), a short novel originally published as a story in 1942, laments the encroachment of civilization on the wild Delta landscape. Another fascinating and recent publication on the Delta region is *This Delta, This Land* by Mikko Saikku (2005). Saikku examines the human-environmental history of the area and especially the prehistoric and historic human impacts on the natural environment, including bird life. It is one of the finest examples of environmental history published on the South. One of the more interesting publications related to southern culture, landscape, and the ivory-bill is Fielding Lewis's *Tales of a Louisiana Duck Hunter* (1988). Lewis describes past duck hunts, close calls with snakes and alligators, and his sightings of the ivory-bill. It is required reading for those interested in outdoor culture in Louisiana.

I found several other books useful for their insights into human-bird relations (Bonta 2003; Cokinos 2000; Doughty 1975; Weidensaul 2002). Not typical ornithological studies, these volumes focus on the human fascination with and historical impacts on various species. They are also more than environmental histories, delving into the larger issue of why human cultures find birds fascinating. Some of this fascination, ironically, has been a factor in the extinction of other birds.

References and Further Reading

Abbey, D. G. 1979. *Life in the Atchafalaya swamp.* Lafayette, La.: Lafayette Natural History Museum.

Abernathy, Y., and R. E. Turner. 1987. U.S. forested wetlands: 1940–1980. *Bio-Science* 37:721–77.

Agey, H. N., and G. M. Heinzmann. 1971a. The ivory-billed woodpecker found in central Florida. *Florida Naturalist* 44, no. 3:46–47, 64.

———. 1971b. Ivory-billed woodpeckers in Florida. *Birding* 3:43.

Allen, A. A. 1924. Vacationing with birds. *Bird-Lore* 26:208–13.

———. 1937. Hunting with a microphone: The voices of vanishing birds. *National Geographic* 71:697–706.

———. 1939. Ivory-billed woodpecker. In *Life histories of North American woodpeckers,* edited by A. C. Bent, 1–12. U.S. National Museum Bulletin 174.

Allen, A. A., and P. P. Kellogg. 1937. Recent observations on the ivory-billed woodpecker. *Auk* 54:164–84.

Allen, B. P. and R. R. Sharitz. 1999. Post-hurricane vegetation dynamics in old-growth forests of Congaree Swamp National Monument. In *On the frontiers of conservation: 10th Conference on Research and Resource Management in Parks and on Public Lands,* edited by D. Harmon, 306–12. Asheville, N.C.: George Wright Society.

Allen, J. A. 1990. Establishment of bottomland oak plantations on the Yazoo Wildlife Refuge complex. *Southern Journal of Applied Forestry* 14:206–10.

———. 1992. Cypress-tupelo swamp restoration in southern Louisiana. *Restoration and Management Notes* 10, no. 2:188–89.

———. 1997. Reforestation of bottomland hardwoods and the issue of woody species diversity. *Restoration Ecology* 5:125–34.

Allen, J. A., and V. Burkett. 1997. Bottomland hardwood forest restoration: Overview of techniques, successes and failures. In *Wetlands and Watershed Management: Science Applications and Public Policy,* edited by J. A. Kusler, D. E. Willard, and H. C. Hull Jr., 328–32. Berne, N.Y.: Association of State Wetland Managers.

Allen, J. A., and H. E. Kennedy Jr. 1989. Bottomland hardwood reforestation in the Lower Mississippi Valley. Slidell, La.: U.S. Department of the Interior, Fish and Wildlife Service, National Wetlands Research Center.

Allen, J. A., S. R. Pezeshki, and J. L. Chambers. 1996. Interaction of flooding and salinity stress on bald cypress (*Taxodium distichum*). *Tree Physiology* 16:307–13.

Allen, P. H. 1958. *A tidewater swamp forest and succession after clearcutting.* Durham, N.C.: Master's thesis, Duke University.

Alig, R. J., H. A. Knight, and R. A. Birdsey. 1986. Recent area changes in southern forest ownerships and cover types. U.S. Forest Service, Southern Forest Experiment Station Research Paper. SE-260.

Amacher, G. S., J. Sullivan, L. Shabman, D. Zepp, and D. Grebner. 1998. Reforestation of flooded farmlands: Policy implications from the Mississippi River Delta. *Journal of Forestry* 96:10–17.

Andren, H., and P. Angelstam. 1988. Elevated predation rates as an edge effect in habitat islands: Experimental evidence. *Ecology* 69 (2):544–47.

Askins, R. A. 1981. *Foraging ecology of temperate-zone and tropical woodpeckers.* Ph.D. diss., University of Minnesota.

Audubon, J. J. 1834. *Ornithological biography, Or an account of the habits of the birds of the United States of America; accompanied by descriptions of the objects represented . . . delineations of American scenery and manners.* London: Adam and Charles Black.

———. 1842. *Birds of America.* Vol. 4. Philadelphia: J. B. Chevalier. Reprint, New York: Dover, 1967.

———. 1999. *Writings and drawings.* Washington: Library of America, No. 113.

Aust, W. M., S. H. Schoenholtz, T. W. Zaebst, and B. A. Szabo. 1997. Recovery status of a tupelo-cypress wetland seven years after disturbance: Silvicultural implications. *Forest Ecology and Management* 90:161–69.

Bailey, A. M. 1939. Ivory-billed woodpecker's beak in an Indian grave in Colorado. *Condor* 41:164.

Bailey, H. H. 1925. *The birds of Florida.* Baltimore: Williams and Wilkins.

———. 1927. The ivory-billed woodpecker in Florida. *Oologist* 44:18–20.

Baker, J. H. 1950. Ivory-bills now have sanctuary. *Audubon Magazine* 52: 391–92.

Barber, J. D., E. P. Wiggers, and R. B. Renken. 1998. Nest-site characterization and reproductive success of Mississippi kites in the Mississippi River floodplains. *Journal of Wildlife Management* 62:1373–78.

Barrow, W. C. 1990. Ecology of insectivorous birds in a bottomland hardwood forest. Ph.D. diss., Louisiana State University.

Bartram, W. 1791. *Travels through North and South Carolina, Georgia, East and West Florida, the extensive territories of the Muscogulges or Creek Confederacy, and the country of the Chactaws.* Reprint, New York: Literary Classics of the United States.

Batista, W. B. 1996. Role of hurricane disturbance in the dynamics of the southern mixed hardwood forest: A case study in northern Florida. Ph.D. diss., Louisiana State University.

Battagalia, L. L., J. R. Keough, and D. W. Pritchett. 1995. Early secondary succession in a southeastern U.S. alluvial floodplain. *Journal of Vegetation Science* 6:769–76.

Bean, A. 1986. The great swamp timber harvest remnants of a legendary engineering feat tell the story. *Georgia Forestry* 39, no. 4:5–7.

Beck, R. E. 1994. The movement in the United States to restoration and creation of wetlands. *Natural Resources Journal* 34:781–822.

Bedinger, M. S. 1981. Hydrology of bottomland hardwood forests of the Mississippi Embayment. In *Wetlands of Bottomland Hardwood Forests,* edited by J. R. Clark, and J. Benforado, 161–76. Amsterdam: Elsevier.

Beyer, G. E. 1900. The ivory-billed woodpecker in Louisiana. *Auk* 17:97–99.

Bick, G. H. 1942. Ivory-billed woodpeckers and wild turkeys in Louisiana. *Auk* 59:431–32.

Bird, A. R. 1932. Ivory-bill is still king! *American Forests* 38:634–35, 667.

Black Bear Conservation Committee. 2005. *Black bear management handbook for Louisiana, Mississippi, Southern Arkansas, and East Texas.* Black Bear Conservation Committee, Baton Rouge.

Bonta, M. 2003. *Seven names for the bellbird: Conservation geography in Honduras.* College Station: Texas A&M University.

Bowling, D. R., and R. C. Kellison. 1983. Bottomland hardwood stand development following clear-cutting. *Southern Journal of Applied Forestry* 7:110–16.

Brewster, W. 1881. With the birds on a Florida river. *Bulletin of the Nuttall Ornithological Club* 6:38–44.

Brodkorb, P. 1970. The paleospecies of woodpeckers. *Quaternary Journal of the Florida Academy of Sciences* 33:132–36.

Burdick, D., D. Cushman, R. Hamilton, and J. Gosselink. 1989. Faunal changes and bottomland hardwood forest loss in the Tensas Watershed, Louisiana. *Conservation Biology* 3, no. 3:282–92.

Carter Ewel, K., and H. T. Odum, eds. 2001. *Cypress swamps.* Gainesville: University Press of Florida.

Catesby, M. 1731. *Natural history of Carolina, Florida and the Bahama Islands.* Vol. 1. Self-published: London.

———. 1985. *Catesby's Birds of Colonial America.* Edited by Alan Feduccia. Chapel Hill: University of North Carolina Press.

Christy, B. 1943. The vanishing ivory-bill. *Audubon Magazine* 45, no. 2:99–102.

Cobb, J. C. 1994. *The most southern place on Earth: The Mississippi Delta and the roots of regional identity.* New York: Oxford University Press.

Cokinos, C. 2000. *Hope is the thing with feathers: A personal chronicle of vanished birds.* New York: Warner Books.

Comeaux, M. 1972. *Atchafalaya swamp life: Settlement and folk occupations.* Baton Rouge: Geoscience and Man Publications.

Conner, W. H., and M. Brody. 1989. Rising water levels and the future of southeastern Louisiana swamp forests. *Estuaries* 12:318–23.

Cottam, C., and P. Knappen. 1939. Food of some uncommon North American birds. *Auk* 56:138–69.

Cowdrey, A. E. 1995. *This land, this South: An environmental history.* Lexington: University of Kentucky Press.

Crompton, D. H. 1950. My search for the ivory-billed woodpecker in Florida. *Bulletin of the Massachusetts Audubon Society* 34, no. 6:235–37.

Dahl, T. E. 1990. Wetlands losses in the United States 1780's to 1980's. Washington, D.C.: U.S. Fish and Wildlife Service.

Dahl, T. E., C. E. Johnson, and W. E. Frayer. 1991. Status and trends of wetlands in the conterminous United States, mid-1970's to mid-1980's. Washington, D.C.: U.S. Fish and Wildlife Service.

Demas, C. 2001. *The Atchafalaya Basin: River of trees.* Washington, D.C.: U.S. Department of the Interior.

Dennis, J. V. 1948. A last remnant of ivory-billed woodpeckers in Cuba. *Auk* 65:497–507.

———. 1967. The ivory-bill flies still. *Audubon Magazine* 69 (6):38–44.

———. 1979. The ivory-billed woodpecker (*Campephilus principalis*). *Avicultural Magazine* 85:75–84.

———. 1984. Tale of two woodpeckers. *Living Bird Quarterly* 3:18–21.

———. 1988. *The great cypress swamps.* Baton Rouge: Louisiana State University Press.

Doughty, R. W. 1975. *Feather fashions and bird preservation.* Berkeley and Los Angeles: University of California Press.

Doyle, T. W., B. D. Keeland, L. E. Gorham, and D. J. Johnson. 1995. Structural impact of Hurricane Andrew on the forested wetlands of the Atchafalaya Basin in south Louisiana. *Journal of Coastal Research* 21:354–64.

Ducks Unlimited. 2005. Ducks Unlimited to plant record number of trees in Arkansas. Press release, July 11, 2005.

Eastman, W. H. 1949. Hunting for ivory-bills in the Big Cypress. *Florida Naturalist* 22:79–80.

———. 1958. Ten-year search for the ivory-billed woodpecker. *Atlantic Naturalist* 13:216–28.

Eckelberry, D. 1961. Search for the rare ivorybill. In *Discovery: Great moments in the lives of outstanding naturalists,* edited by J. K. Terres, 195–207. Philadelphia: Lippincott.

Faulkner, W. *Three famous short novels: "Spotted horses," "Old man," "The bear."* New York: Vintage, 1961.

Fitzpatrick, J. W. 2002. Ivory-bill absent from sounds of the bayous. *Birdscope* 16, no. 3:1, 6.

Fitzpatrick, J. W., M. Lammertink, D. Luneau Jr., T. W. Gallagher, B. R. Harrison, G. M. Sparling, K. V. Rosenberg, R. W. Rohrbaugh, E.C.H. Swarthout, P. H. Wrege, S. Barker Swarthout, M. S. Dantzker, R. A. Charif, T. R. Barksdale, J. V. Remsen Jr., S. D. Simon, and D. Zollner. 2005. Ivory-billed woodpecker (*Campephilus principalis*) persists in continental North America. *Science* 308, no. 5727:1460–62. Available online at http://www.sciencemag.org/cgi/content/abstract/1114103.

———. 2006. Clarification about current research on the status of the ivory-billed woodpecker (*Campephilus principalis*) in Arkansas. *Auk* 123, no. 2 (April): 587–93.

Flynn, K., ed. 1996. *Proceedings of the Southern Forested Wetlands Ecology and Management Conference.* Clemson University, Clemson, S.C.

Forsythe, S.W. 1985. The protection of bottomland hardwood wetlands of the Lower Mississippi Valley. *Transactions of the North American Wildlife and Natural Resources Conference* 50:566–72.

Frayer, W. E., M. J. Monahan, D. C. Bowden, and F. A. Graybill. 1983. *Status and trends of wetlands and deepwater habitats in the conterminous United States, 1950's to 1970's.* U.S. Dept. of the Interior, U.S. Fish and Wildlife service report to Congress. Fort Collins: Colorado State University.

Gallagher, T. 2005. *The Grail bird: Hot on the trail of the ivory-billed woodpecker.* New York: Houghton Mifflin.

Gardiner, E. S., and J. M. Oliver. 2004. Restoration of bottomland hardwood forests in the Lower Mississippi Alluvial Valley, USA. In *Restoration of boreal and temperate forests,* edited by J. A. Stanturf and P. Madsen, 235–51. Boca Raton: CRC Press.

Gorman, J., and A. C. Revkin. 2005. Vindication for a bird and its fans. *New York Times,* August 2, 2005.

Goslin, R. 1945. Bird remains from an Indian village site in Ohio. *Wilson Bulletin* 57:131.

Guilfoyle, M. P. 2001. Management of bottomland hardwood forests for non-game bird communities on Corps of Engineers projects. EMRRP Technical Notes Collection (ERDC TN-EMRRP-SI-21), US Army Engineer Research and Development Center, Vicksburg, Miss. http://el.erdc.usace.army.mil/elpubs/pdf/si21.pdf.

Hamel, P. B., and E. R. Buckner. 1998. How far could a squirrel travel in the treetops? A prehistory of the southern forest. In *Transactions of the 63rd North American Wildlife and Natural Resources Conference,* edited by K. G. Wadsworth, 309–15. Washington, D.C.: Wildlife Management Institute.

Hamilton, R. B. 1975. Central southern region. *American Birds* 29:700–705.

Hamilton, R. B., W. C. Barrow Jr., and K. Ouchley. 2005. Old-growth bottomland hardwood forests as bird habitat, implications for contemporary forest management. In *Ecology and management of bottomland hardwood systems: The state of our understanding*, edited by L. H. Fredrickson, S. L. King, and R. M. Kaminski, 373–88. University of Missouri-Columbia, Gaylord Memorial Laboratory Special Publication No. 10.

Hardy, J. W. 1975. A tape recording of a possible ivory-billed woodpecker call. *American Birds* 29:647–51.

Hasbrouck, E. M. 1891. The present status of the ivory-billed woodpecker (*Campephilus principalis*). *Auk* 8:174–86.

Haynes, R. J. 2004. The development of bottomland forest restoration in the Lower Mississippi River alluvial valley. *Ecological Restoration* 22, no. 3: 170–82.

Haynes, R. J., and L. Moore. 1988. Reestablishment of bottomland hardwoods within national wildlife refuges in the Southeast. In *Proceedings of a conference: Increasing our wetland resources*, 95–103. Washington, D.C.: National Wildlife Federation—Corporate Conservation Council.

Hefner, J. M., and J. D. Brown. 1985. Wetland trends in the southeastern United States. *Wetlands* 4:1–11.

Heinrich, G., and C. Welch 1983. A natural history and current conservation projects to save the ivory-billed woodpecker (*Campephilus principalis*). *AAZP Regional Conference Proceedings*, 224–29.

Hill, G. E., D. J. Mennill, B. W. Rolek, T. L. Hicks, and K. A. Swiston. 2006. Evidence suggesting that ivory-billed woodpeckers (*Campephilus principalis*) exist in Florida. *Avian Conservation and Ecology—Écologie et conservation des oiseaux* 1, no. 3:2. Available online at: http://www.ace-eco.org/vol1/iss3/art2/.

Hoose, P. 2004. *The race to save the Lord God bird*. New York: Farrar, Straus and Giroux.

Hoyt, R. D. 1905. Nesting of the ivory-billed woodpecker in Florida. *Warbler* 1:52–55.

Hudson, P. F. 1998. Meandering processes and channel adjustments in the Lower Mississippi River, prior to major human modifications. Ph.D. diss., Louisiana State University.

Jackson, J. A. 1976. How to determine the status of a woodpecker nest. *Living Bird* 15:205–21.

———. 1988. The southeastern pine forest ecosystem and its birds: Past, present, and future. In *Bird Conservation* 3, edited by J. A. Jackson, 119–59. Madison: University of Wisconsin Press.

———. 1989. Past history, habitats, and present status of the ivory-billed woodpecker (*Campephilus principalis*). Report for the U.S. Fish and Wildlife Service.

———. 1991. Will-o'-the-wisp. *Living Bird Quarterly* 10, no. 1:29–32.

———. 1996. Ivory-billed woodpecker. In *Rare and endangered biota of Florida*, edited by J. A. Rodgers Jr., H. W. Kale II, and H. T. Smith, 103–12. Gainesville: University Press of Florida.

———. 2002a. The truth is out there. *Birder's World* 16, no. 3:40–47.

———. 2002b. Ivory-billed woodpecker (*Campephilus principalis*). The birds of North America online, edited by A. Poole. Ithaca: Cornell Laboratory of Ornithology. Available at: http://bna.birds.cornell.edu/BNA/account/Ivory-billed_Woodpecker/.

———. 2004. *In search of the ivory-billed woodpecker.* Washington: Smithsonian Institute Press.

———. 2006. Ivory-billed woodpecker (*Campephilus principalis*): Hope, and the interfaces of science, conservation, and politics. *Auk* 123:1–15.

Jones, R. H. 1997. Status and habitat of big trees in Congaree Swamp National Monument. *Castanea* 62, no. 1:22–31.

Kennedy, H. E., Jr. 1990. Hardwood reforestation in the South: Landowners can benefit from Conservation Reserve Program incentives. Research Note SO-364. U.S. Department of Agriculture, Forest Service, Southern Forest Experiment Station, New Orleans.

Kilham, L. 1979. Courtship and the pair-bond of pileated woodpeckers. *Auk* 96:587–94.

King, S. L. 2006. Restoration and management of bottomland hardwoods. *Louisiana Agriculture.* Available online at: http://www.lsuagcenter.com/en/communications/publications/agmag/Archive/2006/Spring/Restoration+and+Management+of+Bottomland+Hardwoods.htm.

King, S. L., and B. D. Keeland. 1999. Evaluation of reforestation in the lower Mississippi River alluvial valley. *Restoration Ecology* 7, no. 4:348–59.

King, S. L., B. D. Keeland, and J. L. Moore. 1998. Beaver lodge distributions and damage assessments in a forested wetland ecosystem in the southern United States. *Forest Ecology and Management* 108:1–7.

Klopatek, J. M., R. J. Olson, C. J. Emerson, and J. L. Jones. 1979. Land use conflicts with natural vegetation in the United States. *Environmental Conservation* 6:192–200.

Koenig, W. D. 2005. Persistence in adversity: Lessons from the ivory-billed woodpecker. *Bioscience* 55:646–47.

Lammertink, J. M. 1995. No more hope for the ivory-billed woodpecker *Campephilus principalis. Cotinga* 3:45–47.

Lammertink, M., and A. R. Estrada. 1995. Status of the ivory-billed woodpecker *Campephilus principalis* in Cuba: Almost certainly extinct. *Bird Conservation International* 5:53–59.

Laurance, W. F., and E. Yensen. 1991. Predicting the impacts of edge effects in fragmented habitats. *Biological Conservation* 55, no. 1:77–92.

Lewis, F. 1988. *Tales of a Louisiana duck hunter.* Franklin, La.: Little Atakapas Press.

Loftin, R. W. 1991. Ivory-billed woodpeckers reported in Okefenokee Swamp in 1941–42. *Oriole* 56:74–76.

Louisiana Department of Wildlife and Fisheries, Pearl River Wildlife Management Profile. http://www.wlf.louisiana.gov/hunting/wmas/wmas/list.cfm?wmaid=35.

Louisiana State University, Museum of Natural Science. http://www.museum.lsu.edu/~Remsen/IBW.html.

Lowery, G. H. 1935. The ivory-billed woodpecker in Louisiana. *Proceedings of the Louisiana Academy of Sciences* 2:84–86.

———. 1974. *Louisiana birds.* 3rd ed. Baton Rouge: Louisiana State University Press.

Martin, W. H., S. G. Boyce, and A. C. Echternacht, eds. 1993. *Biodiversity of the southeastern United States: Lowland terrestrial communities.* New York: Wiley.

Martinez, J. D., 1991. Salt domes. *American Scientist* 79, no. 5:420–31.

McIlhenny, E. A. 1941. Ivory-billed woodpecker, general notes. *Auk* 58:582–84.

McKinley, D. 1958. Early record for the ivory-billed woodpecker in Kentucky. *Wilson Bulletin* 70:380–81.

McWilliams, W. H., and J. F. Rosson Jr. 1990. Composition and vulnerability of bottomland hardwood forests of the Coastal Plain Province in the south central United States. *Forest Ecology and Management* 33/34:485–501.

Messina, M. G., and W. H. Conner, eds. 1998. *Southern forested wetlands: Ecology and management.* Boca Raton, Fla.: Lewis Publishers/CRC Press.

Moore, G. E. 1949. Elusive ivory-bills. *Bluebird* 16, no. 12:1.

Murphy, J. L., and J. Farrand Jr. 1979. Prehistoric occurrence of the ivory-billed woodpecker (*Campephilus principalis*), Muskingum County, Ohio. *Ohio Journal of Science* 79:22–23.

The Nature Conservancy. 1992. *The forested wetlands of the Mississippi River, an ecosystem in crisis.* Baton Rouge: The Nature Conservancy.

———. 2006. The Big Woods of Arkansas: An imperiled national treasure. http://www.nature.org/ivorybill/habitat/.

Nelson, J. C., and R. E. Sparks. 1998. Forest compositional change at the confluence of the Illinois and Mississippi rivers. *Transactions of the Illinois State Academy of Science* 91 nos. 1 and 2:33–46.

Nevin, David. 1974. The irresistible, elusive allure of the ivory-bill. *Smithsonian* 4, no. 11:72–81.

Newling, C. J. 1990. Restoration of bottomland hardwood forests in the Lower Mississippi Valley. *Restoration and Management Notes* 8:23–28.

Ouchley, K., R. B. Hamilton, W. C. Barrow Jr., and K. Ouchley. 2000. Historic and present-day forest conditions: Implications for bottomland hardwood forest restoration. *Ecological Restoration* 18, no. 1:21–25.

Ownby, T. 1990. *Subduing Satan: Religion, recreation and manhood in the rural South, 1865–1920.* Chapel Hill: University of North Carolina Press.

Parmalee, P. W. 1958. Remains of rare and extinct birds from Illinois Indian sites. *Auk* 75:169–76.

Peterson, R. T. 1988. My greatest birding moment. *Audubon Magazine* 90, no. 2:64.

Pezeshki, S. R., R. D. Delaune, and W. H. Patrick Jr. 1987. Response of bald cypress (*Taxodium distichum* L. var. *distichum*) to increases in flooding salinity in Louisiana's Mississippi River Deltaic Plain. *Wetlands* 7:1–10.

Putnam, J. A., G. M. Furnival, and J. S. McKnight. 1960. *Management and inventory of southern hardwoods.* Agriculture Handbook 181. U.S. Forest Service.

Reuss, M. 2004. *Designing the bayous: The control of water in the Atchafalaya Basin, 1800–1995.* College Station: Texas A&M University Press.

Rhodes, R. 2004. *John James Audubon: The making of an American.* New York: Knopf.

Ridgway, R. 1898. The home of the ivory-bill. *Osprey* 3:35–36.

Royster, C. 1999. *The fabulous history of the Dismal Swamp Company: A story of George Washington's times.* New York: Knopf.

Rudis, V. A. 1995. Regional forest fragmentation effects on bottomland hardwood community types and resource values. *Landscape Ecology* 10:291–307.

Russell, F. 1973. *The Okefenokee Swamp.* New York: Time-Life Books.

Saikku, M. 1996. Down by the riverside: The disappearing bottomland hardwood forest of southeastern North America. *Environment and History* 2, no. 1:77–95.

———. 2001. Home in the Big Forest: Decline of the ivory-billed woodpecker and its habitat in the United States. In *Encountering the past in nature: Essays in environmental history,* edited by Timo Myllyntaus and Mikko Saikku, 94–140. Athens: Ohio University Press.

———. 2005. *This delta, this land: An environmental history of the Yazoo-Mississippi floodplain.* Athens: University of Georgia Press.

Savage, L., D. W. Pritchett, and C. E. Depoe. 1989. Reforestation of a cleared bottomland hardwood area in northeast Louisiana. *Restoration and Management Notes* 7:88.

Schullery, P., ed. 1986. *Theodore Roosevelt: Wilderness writings.* Salt Lake City: Gibbs M. Smith.

Shaw, S. P., and C. G. Fredine. 1956. *Wetlands of the United States: Their extent and their value to waterfowl and other wildlife.* U.S. Fish and Wildlife Service Circ. 39.

Shear, T. H., T. J. Lent, and S. Fraver. 1996. Comparison of restored and mature bottomland hardwood forests of southwestern Kentucky. *Restoration Ecology* 4:111–23.

Shoch, D. T. 2005. Forest management for ivory-billed woodpeckers (*Campephilus principalis*): A case study in managing an uncertainty. *North American Birds* 59:216–21.

Short, L. L. 1985. Last chance for the ivory-bill. *Natural History* 94, no. 8:66–68.

———. 1987. The ivory-bill: A "galvanizing" effect. *World BirdWatch* 9, no. 3:9.

———. 1988. Status and conservation of woodpeckers. In *Bird Conservation* 3, edited by J. A. Jackson, 161–63. Madison: University of Wisconsin Press.

Short, L. L., and J.F.M. Horne. 1986. The ivorybill still lives. *Natural History* 95, no. 7:26–28.

———. 1990. The ivory-billed woodpecker: The costs of specialization. In *Conservation and management of woodpecker populations*, edited by A. Carlson and G. Auln, 93–98. Report 17, Swedish University of Agriculture and Science, Uppsala.

Sibley, D. A., L. R. Bevier, M. A. Patten, and C. S. Elphick. 2006. Comment on "Ivory-billed woodpecker (*Campephilus principalis*) persists in continental North America." *Science* 311, no. 5767:1555.

Sierra Club, Delta Chapter Profile of the Atchafalaya Basin. http://www.louisiana.sierraclub.org/atchafalaya.asp.

Simpson, B. 1990. *The Great Dismal: A Carolinian's swamp memoir.* Chapel Hill: University of North Carolina Press.

Smith, L. M. 1993. Estimated presettlement and current acres of natural plant communities in Louisiana. Baton Rouge: Louisiana Natural Heritage Program, Louisiana Department of Wildlife and Fisheries.

Smith, W. P., P. B. Hamel, and R. P. Ford. 1993. Mississippi alluvial valley forest conversion: Implications for eastern North American avian fauna. *Proceedings Annual Conference Southeastern Association Fish and Wildlife Agencies* 47:460–69.

Stangel, P. 2005. Ivory-billed rediscovered . . . Now what? *WildBird* 20, no. 1:6–7.

Stanturf, J. A., S. H. Schoenholtz, C. J. Schweitzer, and J. P. Shepard. 2001. Achieving restoration success: Myths in bottomland hardwood forests. *Restoration Ecology* 9, no. 2:189–200.

Sternitzke, H. S. 1976. Impact of changing land use on Delta hardwood forests. *Journal of Forestry* 74:25–27.

Stevenson, H. M., and B. H. Anderson. 1994. *The birdlife of Florida.* Gainesville: University Press of Florida.

Stoddard, H. L. 1969. *Memoirs of a naturalist.* Norman: University of Oklahoma Press.

Tanner, J. T. 1941. Three years with the ivory-billed woodpecker, America's rarest bird. *Audubon Magazine* 43, no. 1:5–14.

———. 1942a. *The ivory-billed woodpecker.* Research Report no. 1. New York: National Audubon Society.

———. 1942b. Present status of the ivory-billed woodpecker. *Wilson Bulletin* 54:57–58.

———. 1986. Distribution of tree species in Louisiana bottomland forests. *Castanea* 51, no. 3:168.

Turner, R. E., S. W. Forsythe, and N. J. Craig. 1988. Bottomland hardwood forest land resources of the southeastern United States. In *Wetlands of bottomland hardwood forests*, edited by J. R. Clark and J. Benforado, 13–28. Amsterdam: Elsevier.

Twedt, D. J., and C. R. Loesch. 1999. Forest area and distribution in the Mississippi alluvial valley: Implications for breeding bird conservation. *Journal of Biogeography* 26, no. 6:1215–24.

Twedt, D. J., R. R. Wilson, J. L. Henne-Kerr, D. A. Grosshuesch. 2002. Avian response to bottomland hardwood reforestation: The first ten years. *Restoration Ecology* 10, no. 4: 645–55.

U.S. Department of Agriculture. Apalachicola National Forest Profile. http://www.fs.fed.us/r8/florida/.

U.S. Fish and Wildlife Service. Bayou Teche National Wildlife Refuge Profile. http://www.fws.gov/refuges/profiles/index.cfm?id=43628.

———. Tensas River National Wildlife Refuge Profile. http://www.fws.gov/refuges/profiles/index.cfm?id=43690.

U.S. National Park Service. Big Cypress National Preserve Profile. http://www.nps.gov/bicy.

———. Big Thicket National Preserve Profile. http://www.nps.gov/bith/.

———. Congaree National Park Profile. http://www.nps.gov/cosw/index.htm.

Walters, J. R., and E. L. Crist. 2005. Rediscovering the king of woodpeckers: Exploring the implications. *Avian Conservation and Ecology—Écologie et conservation des oiseaux* 1, no. 1:6. http://www.ace-eco.org/vol1/iss1/art6/.

Watson, C., and J. Koches. 2002. *A survey for Bachman's warbler* (Vermivora bachmanii) *in the Congaree Swamp National Monument, South Carolina.* April/March 2002. Charleston, S.C.: U.S. Fish and Wildlife Service.

Wayne, A. T. 1895. Notes on the birds of the Wacissa and Aucilla river regions of Florida. *Auk* 12:362–67.

———. 1905. A rare plumage of the ivory-billed woodpecker (*Campephilus principalis*). *Auk* 22:414.

Weidensaul, S. 2002. *The ghost with trembling wings: Science, wishful thinking and the search for lost species.* New York: North Point Press.

Wetmore, A. 1943. Evidence of the former occurrence of the ivory-billed woodpecker in Ohio. *Wilson Bulletin* 55:55.

Wharton, C. H., W. M. Kitchens, and T. W. Sipe. 1982. The ecology of bottomland hardwood swamps of the Southeast: A community profile. U.S. Fish and Wildlife Service. FWS/OBS-81/37.

Williams, J. J. 2001. Ivory-billed dreams, ivory-billed reality. *Birding* 33:514–22.

Williams, M. 1982. Clearing the United States forests: Pivotal years 1810–1860. *Journal of Historical Geography* 8:12–28.

Wilson, A. 1811. *American Ornithology.* Vol. 4. Philadelphia.

———. 1983. *Life and letters of Alexander Wilson.* Philadelphia: American Philosophical Society.

Wright, A. H., and A. A. Wright. 1932. The habitats and composition of the vegetation of Okefenokee Swamp, Georgia. *Ecological Monographs* 2:109–232.

Index